An Alternative History Of Time.

P. J. Naughton

Copyright © P.J. Naughton 2024

P.J. Naughton has asserted his right under the Copyright, Designs and Patents Act 1988 to be identified as the author of this work.

This book is sold subject to the condition that it shall not, by way of trade or otherwise, be lent, resold, hired out, or otherwise circulated without the publisher's prior consent in any form of binding or cover other than that in which it is published and without a similar condition, including this condition, being imposed on the subsequent purchaser.

Version 15

(EW Exl Plan 2022/Last Electron Einstein/Order 4)

Dedicated to all the people of tomorrow, our children and our children's children, the people who will inherit our world, that they will be wise and that they may come to understand the universe better than us.

Contents

1. Introduction...1
2. Blackholes Can't Be Completely Black..2
    Throwing A Stone 2
    The Variable Mass Rocket..2
    Photons Of Light 4
    Detecting Black Holes 5
    The Bending Of Space Time 6
    Diagram 1. Light Retarded By Warped Space Time. 7
    Diagram 2. Intense Gravity Bends Space Time More..9
    Diagram 3. Light Bends To Create A Mirage. 10
    Diagram 4. Light Ray Bent Entering A Black Hole. 10
    Diagram 5. Light Ray Bent Leaving A Black Hole..11
    Diagram 6. Warped Space-Time Bends Light..12
3. An Alternative Interpretation Of Hubble's Law...13
    How Can The Deceleration Of Our Expanding Universe Be Constant?............14
    Could Our Universe Ever Be A Black Hole?..15
    Classical Kinetic Energy Of Our Expanding Universe 16
    Diagram 7. Universe Kinetic Energy v Radius – No Relativity. 16
    Kinetic Energy Of Our Universe According To General Relativity 17
    Diagram 8. Relativistic Kinetic Energy of Universe v Radius. 18
    Diagram 9. Microwave Energy Of Universe Varies With Wavelength. 19
4. How Far Can Gamma Rays Have Come From Outer Space? 20
    Diagram 10. Acceleration Due To Gravity From A Black Hole. 20
    Diagram 11. Frequency Of Light Varies Entering Universe. 22
5. Paradoxes And Mysteries 22
    The Generation Of Mass 22
    Diagram 12. How Mass Might Be Generated. 23
    The Energy Paradox 23
    Why Are Some Stars Thought To Be Older Than The Universe? 24
    Aliens – Are We Alone? 24
    Diagram 13. How Life Supporting Planets May Be Distributed. 25
    Echoes of the Big Bang 30
6. Conclusions 31
7. Final Thoughts 32
Appendix 1 – Photon Frequency In Black Hole..33
    Diagram A1.1. A Photon Moving From Black Hole Centre...............................33
Appendix 2 – Speed Of Expansion Of The Universe Decreases With Time. 37
    Diagram A2.1. Hubble - Galaxies Recede According To Distance. 37
Appendix 3 – Could Our Universe Become A Black Hole?....................................41
    Diagram A3.1. A Particle Escaping From A Planet. 41
Appendix 4 – Classical Universe Kinetic Energy v Radius. 43
    Diagram A4.1 Calculating Classical Kinetic Energy Of The Universe................43
Appendix 5 – Relativistic Kinetic Energy Of Universe 47

Diagram A5.1 Calculating The Kinetic Energy Of Universe. 47
Appendix 6 – Energy Of Photons Moving To Centre Of Universe. 51
Diagram A6.1. Photons Travelling To The Centre Of The Universe. 51
Appendix 7 – Variable Density Of The Universe...55
Diagram A7.1 The Density of the universe at distance r from the centre. 55
Average Density of Universe..57
Appendix 8 – Mass Extinctions 59
Diagram A8.1 – Radiation emitted from the centre of the universe. 59
Table A8.1 Major Extinctions 60
Table A8.2 Calculated emissions from centre of universe t_a 61
Graph A8.1 – 'Big Bang' echoes through centre of universe t_a.........................62
Table A8.3 Calculated times t_2 of extinctions caused by 'Big Bang' echoes 63
Appendix 9 – Average Temperature Of The Universe...65
Appendix 10 – Estimating The Initial Mass of the Universe 67
What could make a tiny universe so unstable? 69
Appendix 11 – Observing An Expanding Universe 71
Diagram A11.1 – Observing An Expanding Universe 71
Graph A11.1 – The Observable universe increases as the expansion slows......73
Table A11.1 – Showing the viewable radius of the universe 74

1. Introduction.

I was inspired to write this text after reading a book by Professor Stephen Hawking which is called 'A Brief History of Time.' If you've never come across it, I would certainly recommend that you read it. It is an excellent summary of the quest by humankind to understand the structure of our universe and our place in it. In summary, it describes very clearly the history of the development of space science and describes what the current thinking is and the problems that exist with current theories relating to the cosmology and space science.

There are a number of concepts and ideas described in 'A Brief History Of Time' which I believe might not be entirely correct. Of course, it's quite possible that I'm *entirely* wrong, but I feel these points which I raise here, are worthy of further debate.

They say mathematics is the language of science - as such there is quite a bit of maths in this work. Someone once told Professor Stephen Hawking that every equation he put in his book would reduce his readership by half. In the end he used only one equation, $E = mc^2$ I've used far more equations but I try to explain them as clearly as possible.

This is a theoretical work which is presented purely for discussion purposes. It should not be relied on as being proven, accurate or correct. I'm presenting what I believe in good faith. I offer no guarantees as to its accuracy or reliability and I reserve the right to change my mind at any future date.

Right or wrong, I hope someone finds something useful in this work and I hope you enjoy reading it and gain something from it that is beneficial both to yourself and hopefully to the rest of humanity. Like many others, I've always strived to discover something useful in the world of Physics. Whether I've succeeded, I'll leave others to decide.

I'm certainly not one of those who thinks everything will be known anytime soon. I feel certain there's plenty out there to be discovered. In truth, I believe we've only just scratched the surface.

Best Wishes,
Patrick Naughton

2. Blackholes Can't Be Completely Black

I think there might be a misunderstanding relating to the nature of 'blackholes'. It's generally thought that photons of light cannot escape from the event horizon of a black hole. But this raises the important question, "what becomes of photons of light that are emitted from the surface of a blackhole?" Would they simply disappear having expended all their energy in a futile bid to escape the attraction of the black hole they emanated from, and if so, what would happen to the energy which they started out with? Alternatively, would photons that had been emitted inside black holes (assuming their existence inside blackholes is even possible), would they be locked forever in an endless orbit around their blackhole host or would they be pulled back towards the centre of the blackhole from where they came, or perhaps they might be absorbed in some way?

A simple application of "Newton's Law of Gravitation" suggests that photons of light would always be able to escape from blackholes, irrespective of the mass or radius of the blackhole from which they emanated. The equivalent frequency of each photon would undoubtedly be redshifted as they escaped. Analysis of this redshift could be useful in the future detection of more blackholes, and importantly for indicating where in space blackholes do not exist.

Throwing A Stone

When a stone is thrown upwards from the surface of a planet, the velocity of the stone gradually reduces as the stone increases in height. The kinetic energy of the stone is reduced and is converted into potential energy. As the stone moves away from the surface of the planet, the vertical velocity of the stone diminishes. Once the vertical velocity reaches zero, the stone will stop moving upwards and begin to fall back towards the surface of the planet. The same would be true of a stone thrown from the surface of a blackhole. Since we believe that no object can travel faster than the speed of light, then it is quite clearly feasible that black holes exist from which no object of *fixed mass* could escape *by being thrown from its surface*.

The Variable Mass Rocket

There is no doubt about the existence of blackholes. There's a blackhole at the centre of most, or perhaps all galaxies, including our own Milky Way galaxy. They could in theory be massive and dense enough to prevent any fixed mass object *thrown from their surface* from ever escaping from their gravitational attraction. However, this does not necessarily mean they could stop *everything* from escaping from their gravitational attraction.

The case of a fixed mass, such as a stone, thrown from the surface of a planet is a very special one because in normal circumstances, without further interaction, the stone cannot gain any more additional kinetic energy, more than it started out with once it has been thrown upwards and started its journey moving away from the surface of the planet. Indeed, as the potential energy of a stone increases as it moves away from the surface of a planet, in the absence of any other means to supply additional energy, the kinetic energy of the stone must inevitably reduce. Since the mass of a stone is fixed, the velocity of a stone must reduce once it is thrown and travels upwards away from the ground.

However, the fixed mass certainly does not apply to the case of a space rocket. A rocket could leave the surface of a blackhole at a speed much lower than the speed of light, far below the escape velocity, but could in theory maintain this low constant speed forever. Its speed need not necessarily alter in any way once it begins its journey, as long as it has enough fuel on board. In fact, its speed could remain absolutely constant. Energy from the fuel could be converted into gravitational potential energy without any need for the speed of the rocket to decrease to compensate for the increase in potential energy.

Furthermore, the mass of the rocket would reduce as the fuel were burnt, thus the gravitational effect on the rocket would be reduced both by the reduction in force of gravity as the rocket climbed in height and by the reduction in mass of the rocket. With sufficient fuel on board, the rocket would be able to move away from any black hole forever. It's worth noting that the speed of such a rocket could be constant as it moved away from any black hole and could theoretically be significantly below the speed of light.

One might argue that space-time would be distorted by a black hole, and it certainly appears that this is the case. But spacetime does not bend at a faster rate at greater distances away from a black hole. Indeed, the rate of distortion with distance must reduce as the effects of gravitational force reduce with distance away from the surface of a black hole. Therefore, even given the distortion of spacetime, the rocket should be able to move ever further away from the surface of a blackhole, even though it would not travel in the exact straight vertical line as seen by a distant observer as it would if it undertook such a mission relative to the surface of a planet which caused zero spacetime distortion.

Some would argue that the rocket would travel the same straight line although this straight line would in fact be bent by the distortion of spacetime. This is true, although the important point remains that the rate of distortion of spacetime would decrease as the distance from the centre of the blackhole increased.

Incidentally if a spaceship could ever be launched from the surface (event horizon) of a blackhole, it would be interesting to know which way spacetime would distort its path, given that the rocket was initially launched vertically. Why should its orbit ever distort in favour of any one direction more than any other if the blackhole were perfectly uniform and spherical? We do see a built-in bias for one particular direction or other in other laws of Physics, e.g. Lenz's righthand law describing the generation of electricity, so it is possible one particular direction could be favoured in preference to another in the case of the blackhole distortion of spacetime, but as yet this appears to remain as one of the many unanswered questions about the structure and behaviour of black holes.

Photons Of Light

In many ways photons of light have more in common with the variable mass rocket than they have with any kind of stone. From the many experiments conducted over recent centuries, we know that photons of light always travel through a vacuum at the same speed – the so-called speed of light. This constant fixed value has been measured to a high degree of accuracy and it's widely believed to be the case that this speed remains the same throughout the entire vast expanse of the universe. As photons move away from a gravitational body, they too must gain gravitational potential energy, which in turn, by the principle of conservation of energy means that they too have to lose another form of energy. The kinetic energy of the photons is reduced but it isn't their speed which changes as we know that this remains fixed. Instead, the effective frequency of the photons reduce as does their equivalent mass.

In this way photons are more like the variable mass rocket than a stone of fixed mass. The gravitational effect reduces far more quickly on photons than it would on a stone of fixed mass. Not only does the gravitational force reduce as the photons move away from the centre of a gravitational body but *also* the force decreases as their effective mass also reduces. Appendix 1 shows the simple calculation made to calculate the reduction in frequency of a photon as it moves away from a gravitational body.

The frequency of a photon of light f_x at any point distance R_x from the centre of a Gravitational body is given by;-

(See appendix 1)

$$f_x = \frac{f_o}{\exp\left[(GM/c^2) \times (1/R_o - 1/R_x)\right]} \quad \ldots 1$$

Where f_x is the frequency of the photon at distance R_x from the centre of a gravitational body of mass M.

f_o is the initial frequency of the photon at distance R_o from the centre of the gravitational body (e.g. which could be the surface or event horizon of a black hole.)

G is Newton's Gravitational constant.
M is the mass of the gravitational body e.g. the mass of a planet or a black hole.
c is the speed of light in space.

It is clear from equation 1 that no matter how great the mass of a black hole is and no matter how far a photon moves away from the centre of the black hole, the frequency of the photon will never reduce to zero. This implies that a black hole could never prevent a photon from escaping from its surface no matter how 'powerful' the black hole proved to be.

If proved to be true, then this is an important result because it suggests that light from our universe could never be confined by a gravitational force. (More on this later!)

At an infinite distance from the centre of a Gravitational body the frequency of a photon becomes ...

$$f_\infty = \frac{f_o}{\exp[(GM/c^2 R_o)]} \quad \ldots 2$$

The light from the surface of a black hole would be significantly affected by its journey from the black hole, but the important conclusion here is that it could never be confined completely by the black hole. The frequency of photons would be reduced and in some cases this reduction would be very significant i.e. an extreme red shift would be observed, but nonetheless, the light would still manage to escape.

Detecting Black Holes

We know from Newton's law of Gravitation that a black hole exists if the following condition is met ...

$$R_o \leq \frac{2GM}{c^2} \quad \ldots 3$$

Substituting from equation 3 into equation 2 gives the condition for a black hole ...

$$f_\infty = \frac{f_o}{e^{0.5}} \quad \ldots 4$$

$$f_\infty = 0.6065 \times f_o \quad \ldots 5$$

This result suggests a measure of redshift of frequency more than 39.3% would be expected at infinite distance from a black hole. Of course, redshift can be caused in other ways (for example the relative velocity of a moving object leads to the Doppler effect) so it would not necessarily be the case that redshift of this scale would be definite proof of the presence of a black hole without further qualification and elimination of other possible causes.

It's worth noting at this point what the impact on our lives would be if our Sun suddenly became a black hole. Our Sun currently has a radius of about 6.96×10^8m. If the Sun's radius were for some reason, suddenly to reduce to about 2.953×10^3m without out any significant change in mass, then the Sun would theoretically become a black hole itself (ignoring the effects of the exclusion principle.)

The gravitational impact on us here on Earth caused by the Sun's transition to a black hole (without any resultant change in its mass) would be that there would be *absolutely no change* in the gravitational pull of the Sun on our planet or indeed on ourselves.

Furthermore, the orbits of all the planets would remain unchanged. They would continue moving around the Sun in exactly the same way that they move around the Sun now, before it collapsed into a black hole. There would however be a change in the spectrum of light that arrives here on Earth which would be radiated by the Sun. Effectively, the spectrum of all light from the Sun would be redshifted so that the frequency of any photon leaving the Sun would effectively arrive on Earth at the frequency given by equation 5 i.e. all photons would fall in frequency by almost 40% compared to the frequency they had on leaving the surface of the Sun.

There is of course a small force on the Earth caused by radiation from the Sun being incident on the Earth's surface. Taking the average incident rate of energy on the Sun's surface as 1,400W/m^2 normal to the Sun, currently the force caused by radiation is about 5.968×10^8 Newtons. This compares to an existing average force of about 4.258×10^{22} Newtons caused by the gravitational pull of the Sun on Earth. Therefore, the force caused by the Sun's radiation on the Earth is less than 1 part in 10^{13} compared to the Sun's gravitational force, so any reduction in the force of radiation would be negligible in terms of changing the Earth's or any other planet's orbit.

There is an existing redshift of light from the Sun, caused by its current gravitational pull, however with its current dimensions, photons from the Sun's surface lose a mere 0.000212% of their frequency on their journey from the surface of the Sun on their way to infinity.

Incidentally, the classical radius of an electron is given as 2.8179×10^{-15}m. We now believe this has little to do with the actual size of an electron, however if the radius of an electron could be reduced to about 1.3526×10^{-57}m whilst maintaining its standard rest mass, then an electron too could theoretically be a black hole (again ignoring the exclusion principle) which illustrates that even particles with tiny mass could theoretically be transformed to become black holes.

So as a first conclusion, the simple application of Newton's law of Gravitation suggests that photons of light should always be able to escape from a black hole regardless of its dimensions of either mass or radius. The light escaping from a black hole will be significantly redshifted and for a body to have dimensions of a black hole, the light escaping from the surface of a black hole must be red shifted by at least 39.3% of its original frequency by the time it reaches an infinite distance from the centre of the black hole from which it originated. This conclusion is based on a relatively simple calculation.

The Bending Of Space Time

Of course, in the previous calculation which was performed to deduce the change in frequency of photons of light emitted from a black hole, no account was made for the bending of spacetime. Einstein's theories of relativity suggest that the bending of spacetime would prevent the photons of light from ever leaving the event horizon of a black hole but I have doubts about this and believe there might be an inconsistency in the theory of relativity in this regard. Perhaps this is just a lack of ability on my part to imagine how spacetime is distorted, but I will attempt to explain my reservations.

I think most people would agree that it's difficult to imagine how spacetime bends exactly. It's hard enough to think of the geometry of everyday objects even in three dimensions of space, then we have to add the fourth dimension of time, then somehow, we have to imagine the resulting spacetime bending in some way. The analogy that is often given to help in this regard is to imagine a marble on a rubber sheet – the rubber sheet is bent by the weight of the marble and this represents the bending of spacetime near a massive gravitational object. This is only a three-dimensional model but at least it helps to have something that we can actually observe, even if it might not be entirely accurate in its depiction (after all it's only at best, a three-dimensional transformation). I think it helps in one regard though, the crucial point to note is that the *rate of bending of spacetime **decreases** as the point of reference moves further away from the centre of the gravitational object in question*.

The argument proposed for the inability of light to escape a black hole is exactly the opposite – that the bending of spacetime would increase as a photon moves away from the surface of a black body, until the bend is so great that light can not escape. Whilst the bending of spacetime may increase as the photon moves away from the surface of a black body, *the rate of bending* (as seen clearly in our rubber sheet analogy) clearly does not.

The same doubts can perhaps be more easily explained by a simple thought experiment. Let us assume that we have a large collapsed star which is close to becoming a black hole. Let us assume we can add mass to this gravitational body without changing its radius in any way. Eventually an event horizon should in theory occur at infinity and as we add more and more mass (without changing the radius of the body), the event horizon should move nearer to the surface of our newly formed black hole. The diagram below illustrates this experiment.

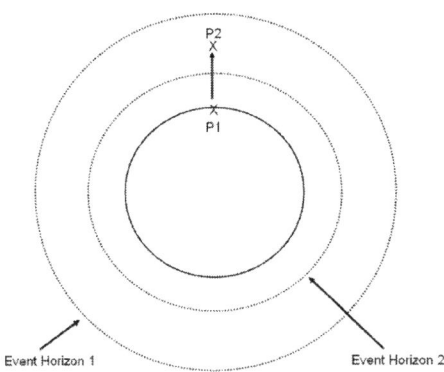

Diagram 1. Light Retarded By Warped Space Time.

Suppose a light is flashed at point P1 at a steady rate. Assume that an observer is at point P2 and that the event horizon is "Event Horizon 1". The light will arrive at point P2. As we increase the mass of our black hole (which we'll assume for the purposes of this experiment that we can do without increasing its radius) the event horizon would gradually move inwards.

Eventually the event horizon would reach "Event Horizon 2". What will the observer at point P2 observe? There are seemingly two contradictory possibilities. The central assumption upon which the theories of relativity are based is that to light will travel at the same speed to all observers independent of their frame of reference. You might therefore expect that the observer at point P2 would continue to see the light flashing at the same constant frequency relative to the change in the pace at which local time was caused to tick resulting from the increase in the pull of gravity.

An alternative possibility is that as the space-time between points P1 and P2 is increasingly distorted by the increasing mass of the black hole, then the frequency of the flashing light would gradually decrease until the event horizon moved nearer to the surface of the black hole than point P2 and then the light would take an infinite amount of time to reach point P2 and the flashing would stop.

Could both possibilities be true? Well in special cases it might appear so. As space-time bends, point P2 could be moved so that it is always within the event horizon so light would always appear to travel towards it at a constant speed, but the straight path could be so warped by the black hole that either the path to it becomes infinite or time slows to such a point that it never arrives. However, there are troubling inconsistencies in this. To say the least, it does involve stretching things a bit. How could the bending of spacetime become infinite at some point away from the surface of the black hole when we know from our rubber sheet analogy that the bending of spacetime decreases as we move away from the black hole?

In any case, even if the bending of space-time caused point P2 to stay within the event horizon, in some cases it could not surely be true for all possibilities for point P2, otherwise the entire universe would always be inside a single black hole. (For example, there is nothing special about P2 to say that we ourselves aren't at a position like this relative to the black hole at the centre of our galaxy).

I believe the concept of an event horizon does have some validity. I believe it marks the boundary around a black hole beyond which it would be impossible *to throw a stone from the surface of the black hole*, but in relation to photons of light or spaceships with plenty of fuel on board, I believe it has more questionable validity.

Both commonsense and experimental evidence suggest that spacetime is bent by large gravitational force. We might therefore expect that the frequency of the flashing light in our experiment would slow down, but I believe it would never reach zero because I do not believe the rate of bending of spacetime could ever become infinite except in one very special case which I'll leave until later to describe.

My understanding of the nature of the bending of spacetime can perhaps be best illustrated by the experiences of a friend of mine, who happens to be a local farmer. Sick of the constant attention of foxes and the theft of so many of his hens, this particular farmer decided to dedicate a fortnight of time doing nothing else but shooting all the many foxes on his farm. He set about this task with real vigour, hunting throughout the night for two whole weeks. A friend of his drove his pickup truck around the fields in the dark, whilst another friend tried to locate the foxes with a spotlight fixed to the back of the vehicle. The farmer fired relentlessly from the trailer at any poor unfortunate fox that happened to be caught in the beam of light.

After a fortnight of hunting, the three of them had killed over two hundred foxes (far more than expected.) However, despite all their efforts, there were just as many foxes on the farm at the end of the two weeks as there had been at the beginning. They came to the conclusion that no matter how many foxes they shot, there'd always be the same number of foxes on the farm. All that was happening was that as they shot a fox, they created a prime vacancy on the farm which meant another fox rapidly moved in from a neighbouring location to fill the vacancy. They would have had to shoot every fox in the country to save their hens by shooting foxes in this way. (I like foxes and I don't harm them, but I understand why poultry farmers try to protect their livelihood!)

I believe that the same sort of effect occurs with spacetime. Basically, the nearer a point is to the surface of a black hole, the more the spacetime is compressed by gravitational forces. I believe this to be a steady and gradual increase as illustrated in the diagram below. As such, spacetime is most compressed at the surface of the black hole P1 and that the compression of spacetime reduces as the point of reference moves away from the surface of the black hole say towards point P2 and beyond.

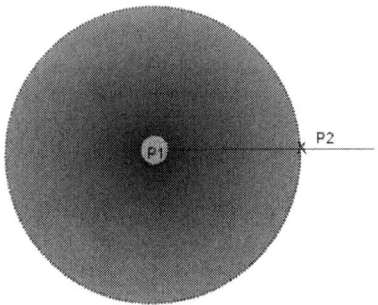

Diagram 2. Intense Gravity Bends Space Time More.

I've mentioned the apparent inconsistency of relativity in predicting constant speeds of light relative to any reference point and the supposed ability of event horizons to mark the boundary of light emitted from a black hole. I've also put forward the argument about the rate of bending of spacetime to be greatest at the point nearest to a black hole to illustrate that an event horizon could not exist for light.

My third argument against the ability of an event horizon to mark the boundary of light emitted from a black hole is to consider what happens to a beam of light as it moves towards a black hole.

I'm sure we are all familiar with the concept of how a mirage is formed. On a warm, still day, the layer of air just above a road surface heats up. The speed of light in hot air is faster than in cold air, so light from the sky tends to curve upwards away from the road surface. To a distant observer looking at the distant road surface, it can appear as though a patch of sky is sitting on the tarmac and this generally gives the appearance of an image of a pool of water.

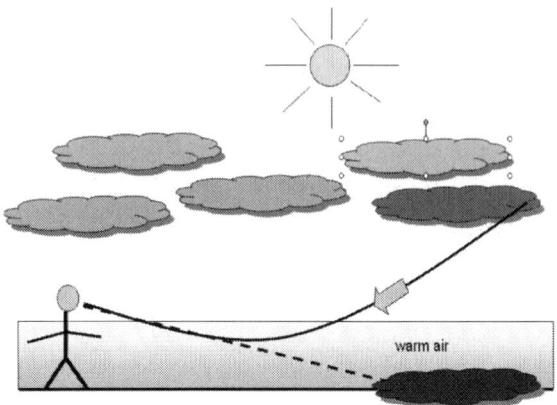

Diagram 3. Light Bends To Create A Mirage.

From our thought experiment, we concluded that in compressed spacetime, unlike in warm air, the speed of light was not increased but instead was diminished. Therefore, the path of a light ray travelling into a black hole is likely to bend in the opposite direction to the rays of light in a mirage situation.

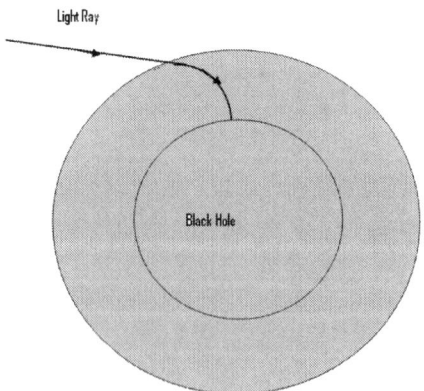

Diagram 4. Light Ray Bent Entering A Black Hole.

Here a ray of light is depicted entering the event horizon of a black hole and being bent inwards towards its surface. Now in every other situation we know that light is bi-directional. By that I mean that if we see a ray of light traced out on a piece of paper, we can find out exactly what the path of the ray of light would be in the opposite direction, simply by reversing the arrows on the ray of light. This is true whether the ray of light is undergoing total internal reflection in a glass prism or is being refracted through a glass lens. There is no reason to suppose the same would not apply to a black hole since this condition holds everywhere else. In that case we can easily see the path a light ray would follow as it left a black hole.

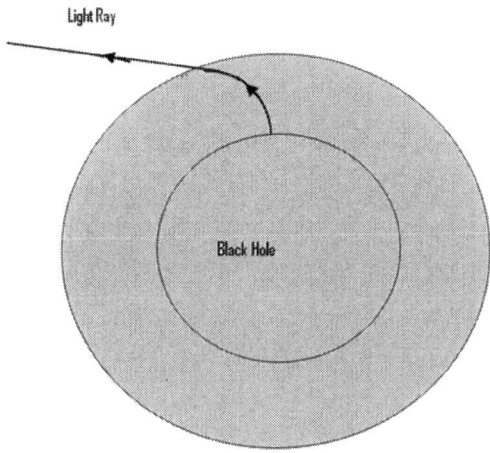

Diagram 5. Light Ray Bent Leaving A Black Hole.

In a way, the central condition upon which General Relativity is based, is possibly an indication of perhaps one of its major short comings. It is based on the assumption that to all observers *in space*, light is seen to travel at a constant speed irrespective of the observer's frame of reference.

Well, we know from Snell's law and the observed refraction of light in a glass lens, that light does not travel at the same speed *in glass* as it does in space. If the path of a beam of light moving away from the centre of a black hole is bent as observed by a distant observer as shown in diagram 5 above, then spacetime is effectively behaving like glass or a layer of warm air in the case of the mirage when it bends light in this way, which we know results from a change in the speed at which light travels. This might suggest that the speed of light is not constant in space, especially in the case when the spacetime through which it travels is highly distorted by gravitation.

Alternative History Of Time

The counter argument is that, in the case of distorted spacetime, the ray of light would not be slowed down but would continue to travel at the speed of light. Only in the case where the distorted spacetime is observed from a distance would the light ray appear to bend, although in reality it continues at the same speed in a straight line relative to the spacetime through which it travels. The central idea here would be that local time would be slowed by the distortion of spacetime, so that as viewed by a local observer, the speed of light would be maintained at the universal constant speed c. However, according to Einstein relativity, this must also be true of a distant observer who might also observe light moving in the same high gravitational field. Their time would not necessarily be slowed to the same extent by the distortion of spacetime if they were far enough away. This seems likely to create an inconsistency. Consider for example the diagram below.

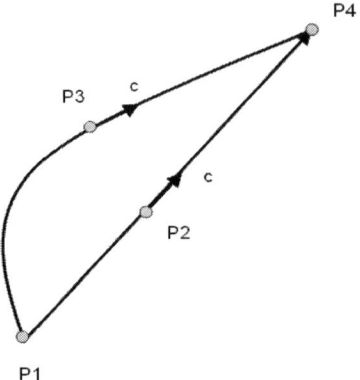

Diagram 6. Warped Space-Time Bends Light.

In a low gravity field, light would travel from **P1** to **P4** via **P2** in a straight line at speed **c**. If spacetime were highly distorted by gravity, the same light ray might travel from **P1** to **P4** via **P3** as observed by a distant observer. The local observer would still see the light travelling in a straight line but a distant observer would see the light travelling in a distorted fashion. The distorted spacetime would slow time for a local observer but it would not slow time for the remote observer as they would not be in the high gravity field, so it seems inconsistent that the speed of light **c** would be observed by the distant observer given that the distant observer would see the light travel in the straight line along path P2 if the gravitational field suddenly reduced to zero.

In summary, the argument here is that the length of the path the light beam is seen to travel must be a different length to the distant observer when they observe a light moving between two points in a high gravitational field compared to the distance travelled by the same light beam travelling between the same two points if the local gravitational field were suddenly reduced to zero, but in both cases the distant observer could be so far away that the rate at which time ticks for them is not effected by the high gravitational field.

Therefore, it would appear inconsistent to claim that the distant observer and the local observer would both see light move at the same speed in these two different cases.

3. An Alternative Interpretation Of Hubble's Law

Whether or not you have accepted any, all, or *absolutely nothing* of what has been stated in chapter 2, this may not necessarily influence how much of my alternative interpretation of Hubble's Law you do or do not agree with.

Through the measurement of the redshift of light, Hubble concluded that most other galaxies are moving away from us. More importantly he found that the speed at which some galaxies move away from us is directly proportional to their distance. This finding was thought by most to be proof that space is slowly expanding but at an ever-increasing rate. However, I believe there might be an alternative interpretation which might explain the findings which Hubble made. If you accept that the universe started from a single 'Big Bang' then the discovery that distant galaxies are moving away from us faster than near by galaxies comes as no real surprise. This is exactly what you would expect to see.

Basically, *the further away you are looking in space, the further back in time you are looking*. Obviously, as is the case in any explosion, particles will move faster nearer to the precise point at which the explosion happened to occur. Hubble's law demonstrates a linear relationship between speed of regression of galaxies and their distance. However, since the light from the galaxies travels at the fixed speed of light c, we can say that *the speed of regression of galaxies is directly proportional to the point back in time that we are receiving light from them*. In other words, the speed of galaxies has reduced *directly proportional to time*. This might possibly indicate that the rate of *deceleration* of the universe is constant. This would be a very special situation but it might prove to be feasible. The possible rate of deceleration of the universe is calculated in appendix 2.

From these simple calculations, we can deduce that the deceleration "a" of the universe has been constant at a value of;-

$$a = -kc \qquad \ldots 6$$

Where **c** is the speed of light and **k** is Hubble's constant. Substituting values into equation 6=>

$$a = -6.8144 \times 10^{-10} \text{ m/s}^2$$

We also derive that the maximum radius of our universe will be;-

$$r_{max} = 6.98 \text{ billion light years} \qquad \ldots 7$$

This equates to a value of **6.59442×10^{25} m**

Alternative History Of Time

How Can The Deceleration Of Our Expanding Universe Be Constant?

If the expansion of our universe is decelerating in a constant way as I have derived, then we can deduce some interesting characteristics of the universe.

According to Newton's Law of Gravitation, the force **F** of attraction caused by gravity between two bodies is given by ;-

$$F = -\frac{GMm}{r^2} \qquad \ldots 8$$

where **G** is Newton's Gravitational constant and **M** and **m** are the masses of two bodies and **r** is the distance between their centres of gravity. The –ve sign indicates the force is attractive.

Force and acceleration are related by Newtons second law of motion;-

$$F = ma \qquad \ldots 9$$

Equation 8 and equation 9 =>

$$a = -\frac{GM}{r^2} \qquad \ldots 10$$

Equation 6 =>

$$a = -kc$$

Therefore, from equation 6 and equation 10 we can deduce;-

$$M = \frac{kcr^2}{G} \qquad \ldots 11$$

Substituting for the maximum value of **r** from equation 7 we can deduce that the maximum value for the mass of our universe would be;-

$$M = 4.44 \times 10^{52} \text{ kg} \qquad \ldots 12$$

Equation 11 suggests the mass of the universe **M** is directly proportional to the radius of the universe *squared*. In other words, the mass of the universe is directly proportional to its surface area. It seems strange on first consideration that this could be possible but later on I will present a simple possible explanation for this phenomenon.

Could Our Universe Ever Be A Black Hole?

The suggestion that the mass of the universe **M** is directly proportional to the radius of the universe squared **r²** raises an interesting question. Could our universe ever be a black hole? The calculation which answers this important question is presented in appendix 3.

From this simple calculation it would appear that our universe will become a black hole at exactly the point at which it reaches its maximum possible radius. Therefore, if you've ever wondered what it might be like to live in a black hole then look around you – so tell us how does it feel? Actually, we are not quite in a black hole *yet* – but I believe we are in a universe that is well on its way to becoming a black hole. But how can we tell that the universe has not already reached its maximum radius and is now contracting again? For a while I thought that might have been a possibility but calculations that follow will suggest that this can not be the case.

It is generally thought (mistakenly) that black holes always have to have some horrifically high acceleration due to gravity but calculations of the value **g** of our universe at its maximum expansion illustrate this is not the case. For comparison the acceleration due to gravity that we experience on the Earth's surface is;-

$$g_E = 9.81 \text{ m/s}^2$$

By way of comparison the acceleration due to gravity caused by the Sun on our Earth's surface is;-

$$g_S = 6.9 \times 10^{-10} \text{ m/s}^2$$

We know from Newton's law of Gravitation that...

$$g = \frac{GM}{r^2} \qquad \ldots 13$$

Substituting in values for maximum radius **r** of the universe and maximum mass **M** from equation 7 and 12 respectively we can calculate the acceleration due to gravity of the universe at maximum size g_U to be;-

$$g_U = 5.94 \times 10^{-11} \text{ m/s}^2 \qquad \ldots 14$$

Clearly the universe at its full expansion would have a huge mass but when you take into account its vast radius, the acceleration due to gravity it can generate at its maximum size is less than a tenth of the pull of the Sun on us here on Earth and we certainly don't feel that in every day life as it is more than 10,000,000,000 times less than the gravitational attraction from our own planet down here on the Earth's surface.

Classical Kinetic Energy Of Our Expanding Universe

This novel interpretation of Hubble's law leads to very interesting predictions for the future behaviour of our universe. I was interested to see how the total kinetic energy **E** of our universe might change with its expansion to see if we could gain any insight in to the way things might turn out. After all there are many differing possibilities. Perhaps the universe could reach a maximum radius and would then start contracting. If that were to happen would time start going backwards? Would we all rise back up from the dead as old people and grow younger until we became children again? First of all, I wanted to see how the kinetic energy of the universe would change with its radius. **Appendix 4** shows the classical calculation I made to calculate the change in kinetic energy **E** of the universe as it expanded.

From appendix 4 we derive a classical prediction for how the total kinetic energy **E** of the universe changes with a change in its radius R_o.

Equation A4.14 =>

$$E = \frac{3kc \times R_o^2 (c^2 - 2kcR_o)}{8G} \qquad \ldots 15$$

If we plot the graph to see how this classical calculation of kinetic energy E varies with the radius of the universe Ro we get the following graph;-

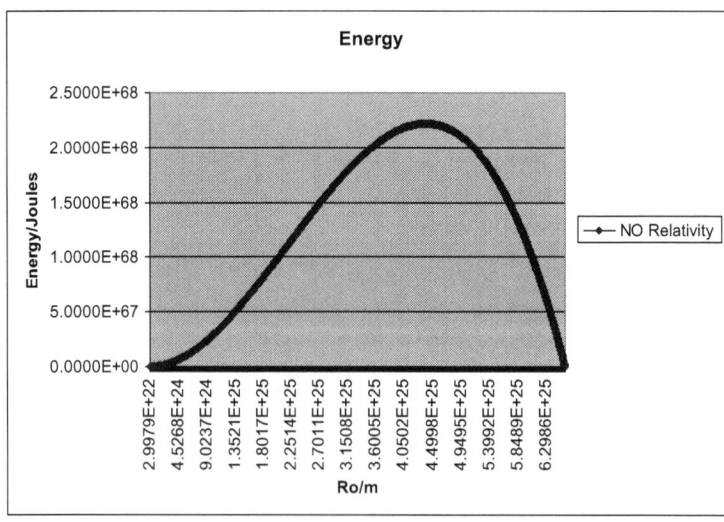

Diagram 7. Universe Kinetic Energy v Radius – No Relativity.

This is an interesting graph in a number of respects. At the maximum radius of the universe that was previously calculated it shows that the total kinetic energy of the universe to be zero. This might suggest that the temperature of the universe is absolute zero unless there is some way that matter with energy could be balanced by matter (or anti matter) with an equivalent amount of negative energy.

In carrying out this calculation I have totally ignored the effect of relativity. Since I am assuming that the initial expansion of the universe would be at the speed of light 'c' it seems *invalid* that the effects of relativity should be ignored. Re-working this same calculation with the effect of relativity taken into account reveals some interesting results.

Kinetic Energy Of Our Universe According To General Relativity

The kinetic energy of an object is given by relativity as;-

$$E = \frac{mc^2}{(1 - v^2/c^2)^{1/2}} - mc^2 \qquad \ldots 16$$

At speeds much below the speed of light this reduces to the more familiar classical expression;-

$$E = \tfrac{1}{2} mv^2 \qquad \ldots 17$$

However, since I am assuming the initial expansion of the universe occurred at the speed of light 'c' it is important to take into account any effects of general relativity. The kinetic energy of the expanding universe with reference to the effects of relativity is calculated in appendix 5.

The results of the calculation in appendix 5 do look slightly complicated, however when the total kinetic energy of the expanding universe with relativity taken into account is plotted on a graph, we do get some startling results. The graph is shown below.

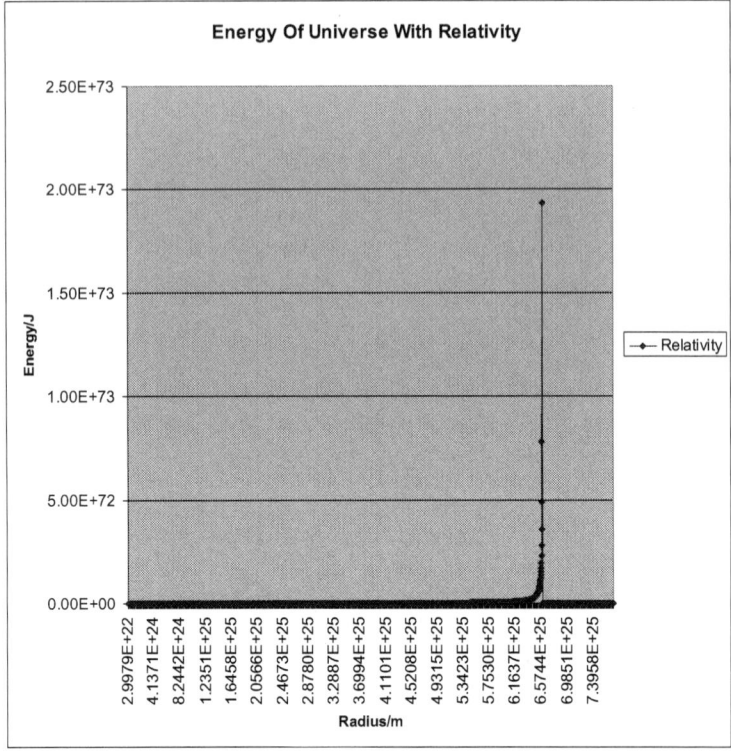

Diagram 8. Relativistic Kinetic Energy of Universe v Radius.

As was the case in the calculation of the total kinetic energy of the expanding universe without taking relativity into account, the calculation with relativity predicts the exact same radius at which the kinetic energy of the universe drops to zero ...

i.e. **R_0 = 6.59442 x 10^{25} m**

However, the graph derived from the relativistic calculation is different from the non-relativistic calculation in a number of important respects. Most notably, at the maximum radius of the universe, relativity predicts that its total kinetic energy will rapidly shoot up to infinity. This would of course have a dramatic impact on the entire universe. Having a finite mass, (albeit a very large one) the temperature of the universe would immediately shoot up to infinity and all matter would be boiled into gases. This graph also shows the total kinetic energy of the universe then plummeting to zero.

The current received wisdom and knowledge causes us to accept relativity as a trusted, reliable and accurate theory. Indeed, many everyday electronic devices would not even function if relativity were not taken into account in their very design. It therefore seems most likely that this graph is the most accurate prediction of what we can expect for the fate of our universe. Until I saw this graph I wondered if the universe could actually be contracting already and how we would know whether it had started contracting or not. Once I saw this graph, I realised if this is correct, we will never experience any slow contraction of the universe.

In these current times there is a great deal of concern about global warming – the heating up of the Earth's atmosphere principally by the trapping of heat from the Sun by carbon dioxide and other gases – often referred to as the green house gases because of the way they could cause heat to be trapped in the Earth's atmosphere in much the same way as the glass of a greenhouse traps heat. I think we are justifiably right to be concerned about global warming. However, the bigger issue which will ultimately lead to the demise of our universe will, I believe, be warming of the universe and I doubt if we will ever be able to do anything about that.

But before I get anyone worried too much about the ultimate demise of our universe let me put your mind at rest. We can easily tell how far our universe has got towards the infinite temperature that my calculation of the effect of relativity on the kinetic energy of the universe predicts. Wein's law tells us that the maximum temperature of a black body T_o and the wavelength of maximum radiation λ_{max} are related by the equation...

$$\lambda_{max} \times T_o = \text{constant} \quad \ldots 18$$

The constant in this case is Wein's constant and it has a value;

Wein's const = 2.896 x 10⁻³ mk

The radiation from the universe has been recorded by many parties over the years (e.g. COBE). The latest measurements suggest that the maximum cosmic radiation occurs at about **λmax = 5.25 x 10⁻² m**

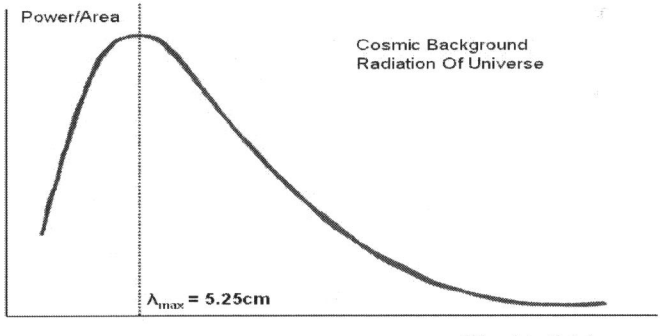

Diagram 9. Microwave Energy Of Universe Varies With Wavelength.

This suggests that the temperature of our universe is still only about
T = 0.0552Kelvin i.e. very, very low and a very long way from reaching infinity.

Actually, measurements of the temperature in space have put the temperature at about **2 Kelvin**. This is still very low. It is possible there are local variations in space, for example due to warming by local stars such as our Sun, or it might be that the temperature is actually higher than our measurements of cosmic radiation suggest. Photons of light travelling from the outer regions of the universe would certainly change energy as they travelled towards its centre. Perhaps this change in energy could be distorting our view of what is really going on at the out reaches of our universe? Could the photons from the outer edges of our universe even reach us? In the next chapter I take a look at what happens to light as it travels towards the centre of our universe from its outer edges.

4. How Far Can Gamma Rays Have Come From Outer Space?

In appendix 1, a calculation shows how photons of light might lose energy as they travel away from a black hole. Conversely, photons will gain energy as they travel towards the centre which would mean their frequency would increase and therefore their wavelength would decease. This would mean that the light we observe coming from the outer edges of our universe will have had a higher wavelength when it set off, which in turn would mean that the temperature of these areas of the universe that we calculate by Wein's law is in reality giving an artificially high result (which itself is very low any way!)

The diagram below shows how the acceleration due to gravity caused by a gravitational body such as a black hole varies with distance from the centre of the gravitational body.

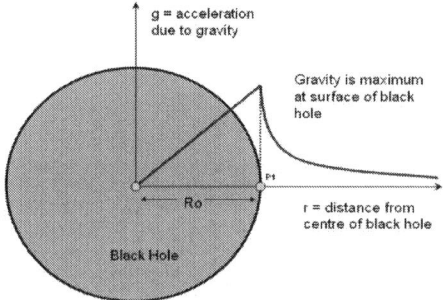

Diagram 10. Acceleration Due To Gravity From A Black Hole.

Clearly the gravitational attraction is greatest at the surface of the gravitational body (this is true whatever the gravitational body happens to be ... a planet, a star, a black hole, our universe, even you sitting in your chair reading this book!)

From the surface of the gravitational body to infinity, the acceleration due to gravity is given by Newton's law of gravitation i.e.

$$g = \frac{-GM}{r^2} \qquad \ldots 19$$

where **G** is Newton's gravitational constant, **M** is the mass of the body and **r** is the distance from the centre of the mass.

Inside the gravitational body the acceleration due to gravity reduces linearly and is given by:-

$$g_r = \frac{-GM}{R_o^2} \times \frac{r}{R_o} \qquad \ldots 20$$

This becomes
$$g_r = \frac{-GM \times r}{R_o^3} \qquad \ldots 21$$

R_o is the radius of the gravitational body.

A photon travelling from infinity to the surface of our universe would clearly gain energy. Although the acceleration due to gravity reduces from the surface of our universe as we move towards its centre, a photon of light would still gain energy as it travels from the outer edge of our universe to the centre of the universe although it would gain energy at a slower rate as the acceleration due to gravity diminishes as it moves closer to the centre.

A calculation is made in Appendix 6 to show how the frequency of a photon of light changes as it moves from the edge of our universe towards the centre. From equation A6.14 it can be seen that at the centre of our universe when it is at its maximum extension then a photon of light leaving the edge of the universe at frequency f_0 would arrive at the centre of the universe at frequency f_1 where...

$$f_1 = f_0 \times 1.6487 \qquad \ldots 22$$

Putting this result together with the result from appendix 1 we can see how the frequency of a photon of light would vary as it moved from infinity into our universe expanded to its maximum size and through to the centre of the universe.

This is a simple calculation, based on several simple assumptions which we'll see later are not entirely correct. Principally it assumes the density of the universe is constant. But if Hubble's constant truly remains fixed and doesn't vary with distance (which may or may not be actually true given the Hubble tension – the discrepancy in measured values) then, if the mass of the universe increases with radius squared then the density of the universe must *increase* towards the centre of the universe.

Alternative History Of Time

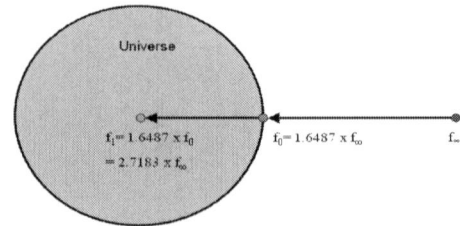

How The Frequency Of Light Varies As A Photon Travels From Infinity Into The Centre Of The Fully Extended Black Hole Universe.

f_1 = frequency of light at the centre of the Universe
f_0 = frequency of light at the edge of the Universe
f_∞ = frequency of light at infinity

Diagram 11. Frequency Of Light Varies Entering Universe.

The frequency of light of a photon travelling from infinity to the centre of the fully expanded universe would increase by 2.7183 times. There would of course be an attendant decrease in the wavelength of the light.

5. Paradoxes And Mysteries

There are a number of paradoxes and mysteries that arise as a result of these calculations.

The Generation Of Mass

How can more mass be generated as the universe expands and why is it that the mass generated increases directly proportionally to the square of the radius of the universe? This does seem a strange concept but we know from quantum mechanics that space is never really empty. Matter and antimatter are spontaneously generated in equal and opposite quantities of positive and negative mass.

It's reasonable to expect positive mass to be drawn by gravitational attraction towards the centre of the universe. Likewise, the negative mass might be repelled by the positive mass universe. In both cases the particles would gain energy. A positive mass gaining energy would, according to relativity, gain mass but a negative mass gaining energy and thus gaining positive mass would inevitably lose negative mass and could eventually disappear. This is not an argument I've ever heard before and therefore may not be true but if proved correct, it might be something which could explain the total lack of antimatter in our universe.

One Possible Mechanism To Explain How The Universe Generates Mass.

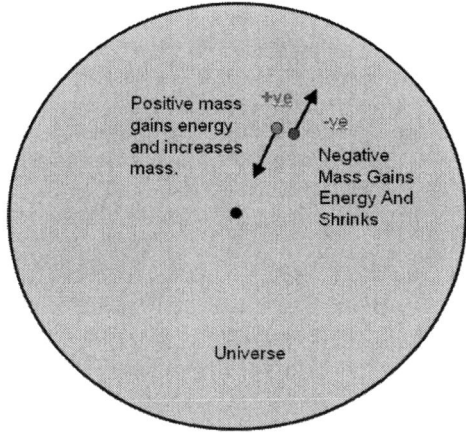

Diagram 12. How Mass Might Be Generated.

This is certainly not the only way in which the universe might generate the prodigious amounts of mass which has clearly been created however it does perhaps illustrate how positive mass could apparently be conjured up out of "nothing".

The Energy Paradox

There is a huge paradox at the heart of Physics relating to the concept of energy which is generally ignored or swept under the carpet. However, it is a very real dilemma. Basically, in order for any object to be raised from the ground we have to give it energy. Suppose we take a brick and lift it one metre above the ground here on Earth, we have to give the brick energy. We have to do work to lift it. Therefore, we say the potential energy of the brick has increased.

If we raise the brick higher, we give it more energy. Suppose we keep on lifting it higher and higher. The gravitational attraction on it gradually reduces but we still have to put energy in to it to move it further away from the ground. Eventually suppose we keep going until we have taken it all the way to infinity. Now the gravitational attraction on the brick is zero. If we let it go what happens? Nothing – the brick will not move because the Earth no longer attracts it. So having given the brick a vast amount of potential energy we suddenly find it hasn't got any energy at all. Physicists generally get over this by saying that the energy level it started with was negative. But this seems inconsistent because the relative potential energy of the brick was zero when it was lying on the ground.

In chapter 3 (diagram 8) we calculated how the kinetic energy of the universe varies with radius and found that at its maximum radius it had both infinite and zero energy. Perhaps this is yet another indication of the same energy paradox that infinite energy is in fact equivalent to zero energy and is perhaps nature's way of ensuring there is no "free lunch". After all, we apparently get a huge amount of mass generated in our universe seemingly out of nothing and perhaps what this infinity-zero duality amounts to is the ultimate pay back.

Why Are Some Stars Thought To Be Older Than The Universe?

Physics is supposed to be about the observation of natural phenomenon. This work does not present any new observational data however it is important that experimental observation should be the final arbiter and that any derived theories should account for all observations in a particular field. It appears that some stars that have been observed are thought to be about 20 billion years old, whilst our universe appears to be only about 13.8 billion years old. This is quite a huge margin of error. How can it be explained?

I think the answer might lie in the same type of error that has repeatedly been made in the past. As human-beings, we have often made the mistake that we are in some ways special. Thoughts of this kind have tended to endure over long periods of our history and have often taken a long time to disprove.

In 340BC Greek philosopher Aristotle postulated that the Earth was stationary and the stars and the Moon and the planets moved in circular orbits around it. In the second century AD Ptolemy added more detail to this model. Not until 1514 did Copernicus propose that *the Sun* was stationary and the Earth and planets moved in circular orbits around their star. Initially this was fiercely resisted but early in the seventeenth century Galileo developed the telescope and the resulting improved observations helped to disprove the stationary nature of the Earth. Subsequently, Newton and Kepler (amongst others) added significantly to the understanding of the mechanics of our universe. The point is, it is less than 500 years since we began to accept, we humans are not actually the centre of the universe.

In the twentieth century, theories from Einstein and Hubble helped us to develop an understanding of how our universe could have been created as a result of a *'Big Bang.'* And that is the prevailing view today – that the *"Big Bang"* initiated the creation of the universe.

But aren't we once again repeating a mistake which our distant forefathers found so hard to dispel? Why should there have been *only one* 'Big Bang'? Aren't we once again making that same old mistake that we are *special* in some way? Surely there could have been hundreds of 'Big Bangs' – possibly thousands. There is no reason why light could not travel between different universes. Indeed, some of the many universes may even overlap and stars from one universe might well be able to move worlds - travelling from the universe in which they were born and moving over to a neighbouring universe. If this turned out to be the case, it is more than possible that our universe could contain stars much older than itself.

Aliens – Are We Alone?

Over the centuries humans have pondered their position in the universe. The eternal question "are we alone?" has often been asked, but to date we have no definite answer either way. However, I think commonsense can be applied to give us at least an indication of our place in the cosmos. As previously stated, in the Dark Ages it was assumed by the Christian church and others that the Earth lay stationary at the centre of the universe and the Sun, the Moon and the stars revolved around us.

As scientific observation progressed, it became patently obvious that this was not really the case at all and that we were in fact a fairly average planet revolving around a fairly average looking star in what, to all intents and purposes, looks like a fairly average galaxy.

If we assume that we are just average, then it's not unfair to assume that the average solar system would be much like ours with a likely average of at least one planet supporting life. Of course, there will be some with more life bearing planets and some with none at all but if we are indeed average (as the current limited evidence suggests) then there are likely to be many planets supporting life – some more advanced than us and some with only primitive life forms.

If we look at our development as humankind, there has clearly been some quite rapid development in the last few thousand years. At the time of Christ, the Romans appear to have been the most advanced people in terms of technology and yet their range didn't extend much more than a few thousand miles. By 1580 Drake had travelled around the world – a journey of some 25,000 miles. By 1969 we had reached the Moon – a round trip of some 500,000 miles. If we continue at this rate of expansion (and there is every evidence to show our rate of technical advance is speeding up rather than slowing down) it's likely that we will increase the distances we can travel by a factor of ~20x every thousand years or so. At this rate of advance we would expect to be able to travel to our nearest star (Alpha Centauri) in about 7 thousand years.

This indeed is a long way off, however if we are indeed about average, it's quite likely that there are other civilisations in our part of the universe in advance of us and if so, they could find us before we find them. Given the three-dimensional nature of space, and assuming more advanced civilisations can travel greater distances than less developed lifeforms, on balance it is likely, that if we are average, there will be civilisations in advance of us who will find us before we find them. Consider the diagram below. If we assume life supporting planets are roughly evenly spaced then in our sector of space there are likely to be eight life supporting planets somewhere around us (in very approximate terms.)

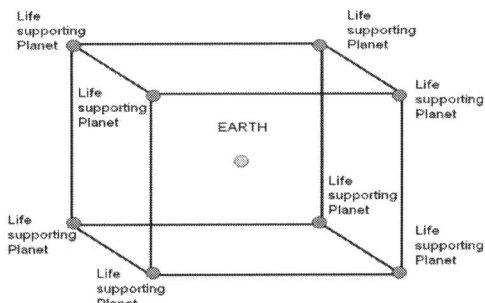

Diagram 13. How Life Supporting Planets May Be Distributed.

It would only take one of these life forms to be in advance of us for the chances to be that we'll be found before we find them. Assuming we are "average" then the chances are at least one of them will find us before we manage to find a planet with *intelligent* life on it. This is of course a very rough approximation based on huge assumptions, not least that intelligent life is evenly spread, but it is perhaps a best possible *guess* in the absence of any other compelling data.

There is some speculation about whether aliens from outer space have ever visited Earth. However, despite many claimed sightings, there appears to be no real firm evidence that any aliens have ever visited Earth. The Roswell incident was certainly peculiar but seems more like an embarrassing undercover experiment by the U.S. military 'gone wrong' rather than a genuine visit by aliens. Some of the Bible stories sound strangely like descriptions of visits by alien beings, but to this day there appears to be no real hard evidence in support of any genuine alien encounter.

On the face of it, the likelihood of any previous human encounter with alien life forms to date sounds unlikely. If there really are aliens out there, surely, we should have seen some of them by now? After all, the cosmos is thought to be some 13.8 billion years old. If some life forms had got to our stage of development, say 5 billion years ago, they would most likely be massively in advance of us by now, assuming that they had not all destroyed themselves. It's possible that some, possibly many alien life forms have indeed become extinct. Afterall, we are aware of many mass extinction events here on Earth. However, some alien civilisations could have survived to reach very advanced stages and could now be able to travel over distances which we currently would find hard to imagine.

So why don't the advanced life forms in space actually visit us? The likely answer is they don't need to, or choose not to do so. If they do actually exist, they are probably able to see quite clearly what is happening on Earth and don't have any reason to come here because there's nothing here that they need. This is the so-called zoo hypothesis. For example, we know there are tigers in the zoo, but we don't really feel the need to try to communicate with them. After all, what would we hope to gain? It'd only be worth the considerable effort required if there was the prospect of any genuine return. Perhaps advanced alien life forms feel much the same way about us?

Any advanced beings are likely to exist a very long way away from us, relative to the distances us human-beings are currently able to travel. I suspect that the conditions in our corner of space have only allowed us to make huge technical advances in very recent times – it was simply not feasible for stone-age men to travel to space and never would have been in their lifetimes, no matter what they had done. The supporting technologies were simply not in existence. However, their endeavours certainly paved the way for more recent advances. One is left to ask though how much more advanced could we have been and how much more advanced could our near neighbours in our part of space really be?

Being able to find a way to travel to our nearest star in about 7,000 years time may seem a very long way in to the future (given our very brief lives) however it will not seem very long compared to the 13.8-billion-year age of our universe. Any alien life form that has developed in advance of us over the life of the cosmos could make a visit by a near neighbour very imminent indeed. In all probability, a visit by aliens would not be a very pleasant experience for residents here on Earth. After all, when we look at the history of Earth, we find that most peoples who were "discovered" by technically more advanced peoples generally faired very badly indeed. Often, they were enslaved, or pushed off their land, or killed off by diseases which they had no immunity to protect themselves against.

If there really were advanced beings in neighbouring solar systems, we might expect to be able to detect them, but in truth this is unlikely just as it would be unlikely for a stone age tribe living in the darkest rain forest to be able to detect our present-day communications by radio or mobile phone. Probably our best defence, maybe our only one, is to make as much technical advance as we possibly can so that we stay ahead of our "space neighbours." However, as always, time could prove to be against us. It's inevitable that sooner or later we will bump into a civilisation much in advance of our own and then our fate will probably be similar to any life forms we have already discovered by then.

Perhaps our best chance is to hope that we learn to leave any life forms we discover alone so that they are able to thrive unmolested in peace and that the life forms in advance of us have already learned the same lesson. It might be the case that the alien life forms currently more advanced than us have already learnt this lesson, which might explain why we haven't met them yet. This is more likely to be the reason why it could be a long time before we meet other life forms not the fact that intergalactic travel looks a bit of a challenge for us mere Earthlings. That would be akin to assuming the Sun rotates around the Earth. Perhaps we should keep reminding ourselves that we as an advanced species are, after all, most likely to be *just average* compared to other so called 'intelligent life' which is not the same as being exactly the same as everybody else of course.

Obviously, there are countless other scenarios to explain why we seemingly haven't encountered any alien life forms yet. One other quite possible explanation is that we really are "first footers", the first intelligent life form to evolve in the universe - after all somebody has to be first, and as unlikely as it may seem, *that might just be us.* There is actually a case to be made by science to suggest this might actually be more than a "distant" possibility. The universe is by nature an extremely violent place. It might not seem like that to us, because as luck would have it, we happen to live in a very quiet neighbourhood. But if we look at the way the universe has developed, we can get some sense of how fortunate we truly are. Immediately after the 'Big Bang' the frontier of space would have spread out very fast indeed. In a later text ('General Relativity Revisited') I calculate that this initial speed of expansion immediately after the 'Big Bang' occurred could have been as high as $1.33c$ where c is the speed of light.

Over the last 13.8byrs the expansion of the universe has gradually slowed down. We can calculate the reduction in this speed from the alternative interpretation of Hubble's law discussed earlier. It turns out that the speed of expansion has decreased by about 2 metres per second per century (incidentally about the same rate the speed of light according to measurements, appears to have speeded up!) That aside, the expansion of the outer boundary of our universe appears to have slowed by about almost c the speed of light over its entire lifetime. If the initial speed of expansion of the universe were so high, the outer boundary would still be moving at a velocity of about $0.33c$ This might not seem very impressive but in terms of temperature, any electron moving at this speed would have a temperature of about 278 million degrees - way too hot to sustain any life form that we can ever imagine.

Of course, the speed of expansion of the inner parts of the universe would be much slower. At a distance r from the centre of the 'Big Bang' the speed of expansion V_r would likely be;

$$V_r = (r/R) \times V_R \qquad ...23$$

where R is the radius of the universe and V_R is the speed of expansion at the outer boundary distance R from the location of the 'Big Bang'. This still means the vast majority of our universe would be far too hot to support life and if correct, it suggests we must be living very much towards the centre of the universe close to the site of the original 'Big Bang'.

We can confirm this in another way. If we look at the density of our neighbourhood in space, we can get a sense of how far we are from the centre of the universe. The gravitational pull of the universe would increase as the mass of the universe increased. The gravitational pull is given by;

$$a = -GM/r^2 \qquad ...24$$

Hubble suggests the deceleration of the universe remains constant at;

$$a = -kc \qquad ...25$$

Therefore, the mass of the universe is given by; -

$$M = kcr^2/G \qquad ...26$$

We can therefore easily calculate the density of the universe at different values of radius r.

$$\Delta M = kcr_2^2/G - kcr_1^2/G \qquad ...27$$

And

$$\Delta V = 4\pi(r_2^3 - r_1^3)/3 \qquad ...28$$

So, the density ρ_r at a distance r from the centre of the universe is given by;

$$\rho_r = \Delta M/\Delta V \qquad ...29$$

It turns out the density of the universe *decreases* dramatically as the distance from the 'Big Bang' increases. If we compare the density in our region of space, we can get a handle on our distance from the location of the 'Big Bang'.

The density of the Inter Galactic Medium IGM is low (1 atom per cubic metre gives a density of about 9.9 x 10^{-27} kg/m^3) but it's the galaxies which reveal the true picture. Roughly speaking, galaxies in our neighbourhood are about 1 million light years apart. Our Milky Way galaxy is seemingly fairly average with a mass of 6 x 10^{42}kg
This gives an average density in our region of space of about;

$$\rho_r = 7.09 \times 10^{-24} \text{ kg/m}^3 \quad \ldots 30$$

This is an incredibly high value and it suggests we are only about 24 million light years from the site of the original 'Big Bang'. This might sound like a very long way but if the initial speed of the expanding universe were u = 1.33c the universe would have a radius of about 11.6 billion light years by now (assuming it started expanding at 1.33c and has been slowing at a steady rate of a = -kc as given by Hubble's law over the last 13.8 billion years). If this were the case then we'd be located about 0.2% from the centre of the universe.

Even at such a relativity short distance to the centre of the universe to the place where the 'Big Bang' originally occurred, our temperature would still be about 653°c - far too hot to support life. So, either the universe did not start off expanding at such a high speed, or the universe is older than we think. Either way it's clear, *only the most central portion of our universe is cool enough to support life and in relative terms, it wasn't long ago that it cooled sufficiently for life to evolve.*

r	M	ΔM	Volume = 4/3πx r^3	ΔV	Density @r	r lyrs
2.2915E+23	5.3620E+47	4.6798E+43	5.04020E+70	6.59828E+66	7.0925E-24	2.4220E+07
2.2916E+23	5.3625E+47	4.6800E+43	5.04086E+70	6.59885E+66	7.0921E-24	2.4222E+07
2.2917E+23	5.3629E+47	4.6802E+43	5.04152E+70	6.59943E+66	7.0918E-24	2.4223E+07
2.2918E+23	5.3634E+47	4.6804E+43	5.04218E+70	6.60001E+66	7.0915E-24	2.4224E+07
2.2919E+23	5.3639E+47	4.6806E+43	5.04284E+70	6.60058E+66	7.0912E-24	2.4225E+07
2.2920E+23	5.3643E+47	4.6808E+43	5.04350E+70	6.60116E+66	7.0909E-24	2.4226E+07
2.2921E+23	5.3648E+47	4.6810E+43	5.04417E+70	6.60173E+66	7.0906E-24	2.4227E+07
2.2922E+23	5.3653E+47	4.6812E+43	5.04483E+70	6.60231E+66	7.0903E-24	2.4228E+07
2.2923E+23	5.3657E+47	4.6814E+43	5.04549E+70	6.60289E+66	7.0900E-24	2.4229E+07
2.2924E+23	5.3662E+47	4.6816E+43	5.04615E+70	6.60346E+66	7.0897E-24	2.4230E+07
2.2925E+23	5.3667E+47	4.6818E+43	5.04681E+70	6.60404E+66	7.0894E-24	2.4231E+07
2.2926E+23	5.3671E+47	4.6820E+43	5.04747E+70	6.60461E+66	7.0891E-24	2.4232E+07

Table 1. How the density of the universe varies with distance from the centre (see appendix 7).

If we repeat this calculation but assume the initial expansion of the universe was just the speed of light i.e. u = c and the same constant Hubble deceleration had occurred, then the outer edge of the universe would by now have reached a radius of 6.969 billion light years and it would have slowed to v = 3.23 million m/s

If this is the case, the universe will eventually stop expanding some 13.95 billion years after the 'Big Bang' some 150 million years from now, having reached a maximum radius of 6.970 billion light years, just one million miles bigger than it is now.

At a distance of 24 million light years from the centre of the universe, our electron temperature in our area of space would be about T = 2.73 Kelvin which is much more in line with measurements that are made. If the outer edge of the universe is presently moving at V = 3.23 million m/s then the electron temperature on the outer edge would be about 229,000 Kelvin (given by $T = mv^2/3K_b$ where K_b is Boltzman's constant). At 24 million light years from the centre, we'd be about 0.34% from the centre of the universe given that the current radius of the universe would be 6.969 billion light years. Any region of space *more than* 265 million light years from the centre of the universe (about 3.8% along the radius from the centre of the universe) would have an electron temperature in excess of 60°c (a temperature thought to be too hot to support most forms of life).

If this calculation regarding our place in the universe is correct, then all those texts from ancient times and the Geocentric theory of Claudius Ptolemy that claimed we are at the centre of the universe *might actually turn out to have a grain of truth in them after all!* (although there's no suggestion here that the sun orbits the Earth!)

And this could be a very compelling argument to explain the lack of *long-distance* space tourists. Perhaps the only other life forms that could find suitable conditions for survival will be in our (in space terms) fairly near neighbourhood, and therefore the chance of finding life in the furthest reaches of the universe at any point in time might be extremely unlikely indeed.

Having reached its maximum size, our universe might begin the process of collapsing in on itself. This process of collapse is likely to take just as long, perhaps (because of complex gravitational structures and the resistance of compressed matter) even slightly longer than the process of expansion. Once it collapses to an extremely small size, with a mass less than a fraction of a single proton (see appendix 10) our universe will eventually become unstable once more and explode in yet another 'Big Bang' and the whole process might possibly start over again and repeat *ad infinitum* in an endless yoyo effect. However, there might be a violation of the uncertainty principle as the universe nears its maximum extent which might interrupt this process (see my later text – 'The Big Bang').

Echoes of the Big Bang

It is possible that echoes from our original 'Big Bang' might have caused subsequent mass extinctions here on Earth (see appendix 8). It is very likely that a massive amount of energy was released at the occurrence of the 'Big Bang'. It is unknown exactly what might have caused this event but it might be the case that it led to a massive release of electromagnetic radiation which could have travelled endlessly, back and forth across the universe, moving forever at the speed of light 'c'. This radiation might continue to repeatedly bounce back and forth off the boundary of the universe to this day, leading to significant impact to any planetary body it might encounter on its journey.

In appendix 8, the timings of recent mass extinctions are used to calculate how this radiation might have travelled across our universe. This hypothesis then provides an indication of the possible timing for future extinction events. There are potentially many possible causes of mass extinctions, but it seems likely that an echo from the 'Big Bang' will not pose any threat to Earth for several hundred million years.

However, it isn't certain that any such event would necessarily lead to the extinction of human life. There are a number of important factors in our favour;

First, we must remember that life continues until this day, after the occurrence of all previous mass extinction events. The last mass extinction event saw the end of the existence of the largest land-dwelling dinosaurs but plenty of life continued including mammals and many forms of sea-life.

Secondly, we should remember that although the last of the great dinosaurs was killed off in the last major extinction, this process appears to have taken about 10,000 years to complete, so if any similar event occurred there would hopefully be time to develop some kind of survival strategy. It is generally thought that this last major extinction event came about because of a meteorite strike. This may well have been the case although it still remains unexplained why if this was the case, this extinction process took so long to fully play out.

Third, we can perhaps have hope that the next extinction event might be less intense than previous such events, given that the universe will inevitably be that much older.

Fourth, and possibly the most important, we humans have already developed impressive technology and the pace of change is ever accelerating. We are no longer completely beholden to the whims of nature and can, to some extent, develop a strategy to protect ourselves. It is most important that leading nations learn to set aside their differences and cooperate with each other in full in order to work out the best strategy to protect all life on Earth in the interests of us all.

It would probably be particularly helpful if we could work out the exact nature of previous extinction events - it's likely that clues will still exist and be there for us to find if we bother to look hard enough to identify them. Knowing the exact nature of the challenge we are likely to face, could very well help us to prepare for any major threat of extinction that might arise at some point in the future.

One of the many final questions which remains is to try to work out exactly what might happen if the universe does, as I suspect, eventually stop expanding. In appendix 11 I've used all the basic assumptions I've reached here to make a simple estimate of the extent to which we can observe the expansion of the boundary of the universe. Having reached its maximum size, it's uncertain whether the universe would begin to contract again and if it did start to fall back in on itself, what impact this might have on the universe as a whole. Would the universe possibly heat up again? And might it be the case that time would start to go backwards or would the universe possibly reach some sort of dormant state instead? Or would a violation of the uncertainty principle as the momentum of the universe reduced to zero, suddenly lead to another 'Big Bang'?

As far as the calculations made here are concerned, all this seems very much in the future – possibly 150 million years away. This might sound like a very long time away in comparison to our short life spans, but in comparison to the age of the universe, this is a very small fraction of time indeed. And given the level of uncertainty surrounding most of the measurements of the universe, we might reach this point at any time – possibly much sooner than we expect. Perhaps one note of comfort is that whatever happens at the boundary of the universe, it is likely to take many billions of years to affect us given the massive size of the universe, so there is hopefully a great deal of time that we can use to further develop our understanding of cosmology to optimize the chances of the ongoing survival of the human race.

6. Conclusions

In summary I'd like to briefly list the most likely outcomes reached in this brief work;-

- Our universe will stop expanding eventually.
- When it reaches its maximum size our universe will become a black hole (just.)
- There are possibly many universes in existence at the same time – each emanating from their own 'Big Bang'.
- Light can escape from 'blackholes'.
- Light can travel between the various universes and as such we should be able to observe them.
- It seems likely we are located very near the centre of our universe at a point fairly close to the place where the 'Big Bang' occurred and any alien life forms that do exist are likely to be located in a similar location.
- It is likely that echoes of the 'Big Bang' have caused mass extinctions here on Earth and may cause future events but probably not for several hundred million years after which time the universe will have stopped expanding.
- Our universe might eventually collapse down to an extremely small size with a mass less than a proton at which point it might experience another 'Big Bang'. The whole process of expansion followed by collapse might repeat ad infinitum unless the violation of the uncertainty principle does not lead to another 'Big Bang' once the expansion of the universe comes to a halt.

7. Final Thoughts

I may not be correct in many of these thoughts; I could well be wrong in all of them. But I think this is probably the fate of us all. I'm certainly not one of these who thinks that any time in the next few years we'll discover an astonishing new equation or a new theory and as a result of that, we'll be able to understand how everything works in the entire universe.

I think being a scientist is a bit like being a record-breaking runner. You can't expect to be the best for all time – the best you can hope for is to be the best for a while.

Even if everything I've written here turns out to be false, I don't really care. However, if it inspires just one person to write a paper or a book to correct anything that I've written, then I'll consider my endeavour a success.

Appendices

Appendix 1 – Photon Frequency In Black Hole

A quantum of light mass "m" moves away from a black hole of mass "M" at a distance "r" from the centre of the black hole. The black hole has radius "R_o"

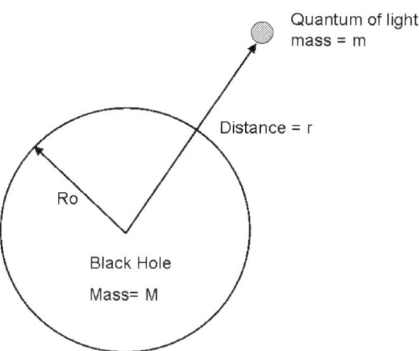

Diagram A1.1. A Photon Moving From Black Hole Centre.

The force F on the quantum of light is given by; -

$$F = \frac{d(mv)}{dt} \qquad \ldots A1.1$$

Where v is the velocity of the photon of light.

Therefore …

$$F = m \times \frac{dv}{dt} + v \times \frac{dm}{dt} \qquad \ldots A1.2$$

For a quantum of light, the velocity v is thought to be constant in space as predicted by Einstein's theory of relativity. i.e. **v = c**

Therefore…

$$\frac{dv}{dt} = 0 \qquad \ldots A1.3$$

Alternative History Of Time

Substituting equation A1.3 into equation A1.2 gives...

$$F = c \times \frac{dm}{dt} \qquad \text{...A1.4}$$

The kinetic energy lost by the photon **dε** in moving distance **dx** further away from a black hole is...

$$d\varepsilon = F\,dx \qquad \text{...A1.5}$$

Where **F** is the gravitational attraction on the photon.

Substituting from equation A1.4 into equation A1.5 and since **dx/dt = c** we get ...

$$d\varepsilon = c^2\,dm \qquad \text{...A1.6}$$

Newton's Law of Gravitation gives the gravitational force on any body of mass m at distance r from the centre of the gravitational body of mass **M** as ...

$$F = \frac{GMm}{r^2} \qquad \text{...A1.7}$$

Where G is Newton's Gravitational constant.

Equations A1.4, A1.5 and A1.6 =>

$$\frac{GMm}{r^2}\,dx = c^2\,dm \qquad \text{...A1.8}$$

The effective mass of the photon **m** varies as the photon loses kinetic energy as it journeys away from a black hole. Rearranging equation A1.8 gives...

$$\frac{GM}{c^2 r^2}\,dx = \frac{1}{m}\,dm \qquad \text{...A1.9}$$

Integrating both sides of equation A1.9 and substituting boundary values we get...

$$\frac{GM}{c^2} \times [1/R_o - 1/R] = \log_e m_o - \log_e m_r \qquad \text{...A1.10}$$

Where m_o is the mass of the photon at distance R_o from the centre of the black hole (i.e. on its surface) and m_r is the mass of the photon at some distance **R** from the centre of the black hole.
Equation A1.10 =>

$$\frac{GM}{c^2} \times [1/R_o - 1/R] = \log_e \frac{m_o}{m_r} \qquad \text{...A1.11}$$

We know from Einstein's famous equation that the total energy of any body is such that....

$$E = mc^2 \quad \text{...A1.12}$$

We also know that the same energy for a photon can be expressed in terms of the frequency of the photon...

$$E = hf \quad \text{...A1.13}$$

Where **h** is Planck's constant and **f** is the equivalent frequency of the packet of light.

Equations A1.12 and A1.13 =>

The equivalent "mass" of a photon at any point is...

$$m = \frac{hf}{c^2}$$

Substituting this into equation A1.11 we can express the equivalent frequency of a photon of light at any point as it travels away from a black hole ...

$$f_x = \frac{f_o}{\exp[(GM/c^2) \times (1/R_o - 1/R_x)]} \quad \text{... A1.14}$$

Where the equivalent frequency of the packet of light f_x at any point R_x from the centre of a black body is related to the frequency of the photon f_o emitted from the surface of a black body at a distance R_o from the centre of the black body.

Alternative History Of Time

Appendix 2 – Speed Of Expansion Of The Universe Decreases With Time.

Hubble's law tells us that the further away a galaxy is located, the faster it appears to be moving away from us. The velocity at which a galaxy moves relative to ourselves V is directly proportionate to the distance **x** that the galaxy appears to be from us. In simple terms...

$$V = kx \qquad \ldots A2.1$$

where k is Hubble's constant.

Hubble's constant is generally quoted as;-

$k = 70.1$ km/s/Mpc

1 Mpc = 3,262,000 light years. Therefore, in mks units Hubble's constant becomes ....

$k = 2.27306 \times 10^{-18}$ s^{-1}

This apparent relationship between the velocity at which galaxies move away from us and their distance could be explained by the very concept that we are seeing distant galaxies not as they are now but as they were sometime in the past. The further away they appear to be located, the further into the past we are seeing them. After the "Big Bang" it is probably fair to assume that the initial speed at which matter was flung out of the initial explosion was very high. However, over time this velocity probably reduced as matter was slowed by gravitational force.

One of the most surprising aspects of Hubble's law is the exceptionally simple relationship it gives us between the speed of galaxies relative to ourselves and their distance from us. Clearly this is a linear relationship...

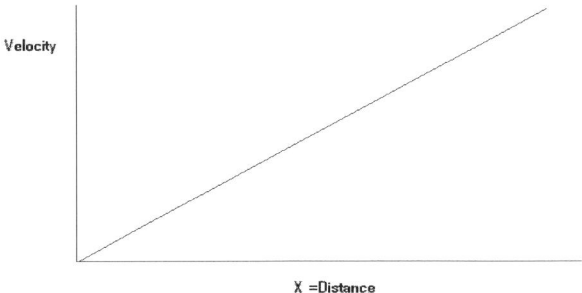

Diagram A2.1. Hubble - Galaxies Recede According To Distance.

However, we know that light travels over distance 'x' at a constant speed 'c' so we can deduce that the relationship between velocity V and time t is linear – in other words Hubble's law is describing linear motion i.e. a constant deceleration for the expansion of the universe after the 'Big Bang.'

From equation A2.1 we know that the speed of regression of a galaxy is given by....

$$V = kx$$

But x = ct (where c is the speed of light and t is the time for that light to travel distance x)

Taking two different galaxies at different distances away from us we get ...

$$V_1 = kct_1 \qquad \text{...A2.2}$$

$$V_2 = kct_2 \qquad \text{...A2.3}$$

Equation A2.2 – equation A2.3 =>

$$V_1 - V_2 = kct_1 - kct_2$$

$$V_1 - V_2 = kc(t_1 - t_2) \qquad \text{...A2.4}$$

However, we know that for linear motion ....

V = U + at (where "U" is the initial speed of the body and "a" is the acceleration.)

Therefore...

$$V_1 = U + at_1 \qquad \text{...A2.5}$$

$$V_2 = U + at_2 \qquad \text{...A2.6}$$

Equation A2.5 – equation A2.6 =>

$$V_1 - V_2 = a(t_1 - t_2) \qquad \text{...A2.7}$$

Comparing equation A2.4 and equation A2.7 we can see that Hubble's Law is giving us an expression for the deceleration of the universe...

$$a = kc \qquad \text{...A2.8}$$

Given that the speed of light c = 2.9979×10^8 m/s

Then from our value of k we can deduce ...

$$a = -6.8144 \times 10^{-10} \text{ m/s}^2 \qquad \text{...A2.9}$$

The minus sign is inserted as an indication of the deceleration from the base point of the "Big Bang"

This value of constant deceleration for the expanding universe allows us to calculate a number of interesting predictions for our universe.

For example, if we were to assume that the maximum speed of matter immediately after the "Big Bang" was "c" the speed of light then....

From the simple formula for constant acceleration, we know that ....

$$V = U + at \qquad \text{...A2.10}$$

At a point in time when the universe stops expanding V = 0

Therefore, equation A2.10 =>

$0 = c - at$

$t = c/a$

Substituting values for c = 2.9979 x10^8 m/s and **a** from equation A2.9 =>

$$t = 13.95 \text{ billion years} \qquad \text{...A2.11}$$

Therefore, the universe will stop expanding 13.95 billion years after the "Big Bang" about 150 million years from now.

Similarly, we can deduce what size the universe will reach by the time it stops expanding...

We know for linear motion ....

$$V^2 = U^2 + 2as \qquad \text{...A2.12}$$

Where s is the distance travelled.
Assuming V = 0 and U = c then ....

$$0 = c^2 + 2as$$

Substituting our value for a from equation A2.9 =>

The maximum radius of the universe will be...

$$r_{max} = 6.98 \text{ billion light years} \qquad \text{...A2.13}$$

i.e. the maximum diameter of the universe will be...

$$d_{max} = 13.95 \text{ billion light years} \qquad ...A2.14$$

There are many more interesting deductions that can be made from these simple calculations. Perhaps the most fascinating will be the relationship between Hubble's Law and Newton's Law of gravitation in helping us to understand how the universe is expanding and the fact that Hubble's Law is predicting a constant deceleration for the expansion of the universe. This is not something that one would necessarily expect from a simple understanding of the law of gravitation and it perhaps tells us something about the way the mass of the universe must be changing.

Appendix 3 – Could Our Universe Become A Black Hole?

Newton's Law of Gravitation tells us that the force F exerted on a mass m at distance r from the centre of a large gravitational body of mass M is given by;-

$$F = \frac{GMm}{r^2} \quad \ldots A3.1$$

Where G is Newton's Gravitational constant.

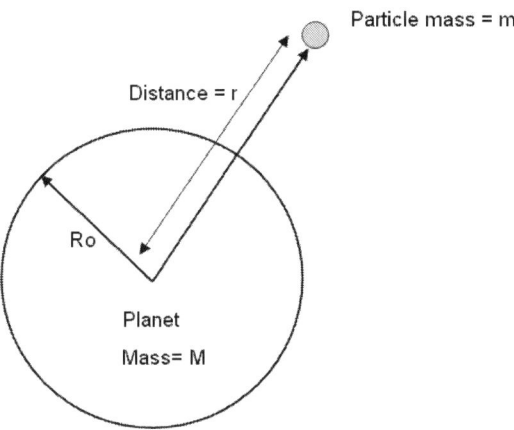

Diagram A3.1. A Particle Escaping From A Planet.

The increase in potential energy dE required for the mass m to move by a small distance dx is given by;-

$$dE = Fdx \quad \ldots A3.2$$

Substituting from equation A3_1 into equation A3_2 and integrating dE we get...

$$E = \int \frac{GMm\, dx}{r^2} \quad \ldots A3.3$$

Completing the integration on the right-hand side gives...

$$E = \left[\frac{GMm}{r} \right]_{R2}^{R1} \quad \ldots A3.4$$

where the mass m moves from $r = R_1$ to $r = R_2$

For mass M to be on the limit of being a black hole then the mass m could just reach infinity if it had the maximum kinetic energy on leaving the surface of the planet i.e. if it were thrown at the maximum speed c.

Therefore...

$$R_2 = \infty$$

Therefore, at the limit of M just being a black hole...

$$E = \frac{GMm}{R1} \qquad \ldots A3.5$$

The maximum kinetic energy of the mass **m** is ...

$$E = \tfrac{1}{2} m c^2 \qquad \ldots A3.6$$

Equations A3.5 and A3.6 =>

$$c^2 = \frac{2GM}{R1} \qquad \ldots A3.7$$

Substituting for the maximum mass of the universe M from equation 12 where we found M_{max}= 4.44E+52 kg then equation A3.7 =>

$$R_1 = 6.5944234965 \times 10^{25} \text{ m for M to be a black hole.}$$

Comparing this with the calculated maximum radius of our universe from equation 7 we get ...

$$R_{max} = 6.59442 \times 10^{25} \text{ m}$$

We see that there is an exact match!!!! This is a very important conclusion. It means when our universe has expanded as far as it can expand, it will have become a black hole.

CONCLUSION:

This means our universe will become a black hole exactly at the point that it reaches its maximum radius.

Appendix 4 – Classical Universe Kinetic Energy v Radius.

Assume the universe mass M is expanding. At any point in time the radius of the universe is Ro.

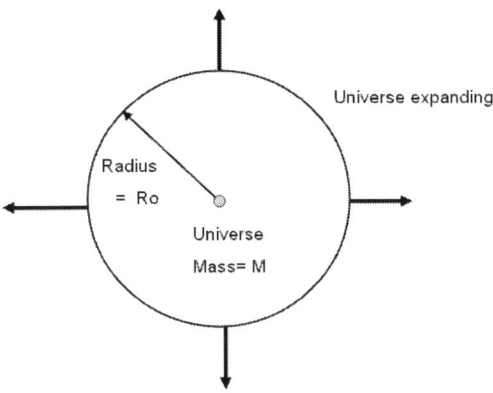

Diagram A4.1 Calculating Classical Kinetic Energy Of The Universe.

For constant acceleration a we know that...

$$V^2 = U^2 + 2aS \qquad \ldots A4.1$$

Where U is the initial velocity and V is the velocity after travelling distance S

If V_o is the velocity of the surface of the universe (at radius R_o) then at any point r ;-

$$V = V_o \times \frac{r}{R_o} \qquad \ldots A4.2$$

We know that the kinetic energy of a moving body is;-

$$E = \tfrac{1}{2} M \times V^2 \qquad \ldots A4.3$$

The mass of the universe can be calculated as;-

$$M = \int 4\pi r^2 \, dr \times \rho \qquad \ldots A4.4$$

where ρ is assumed to be the constant density of the universe.
Equation 11 =>

$$M = \frac{kcr^2}{G} \quad \ldots A4.5$$

$$\rho = M/Vol \quad \ldots A4.6$$

$$Vol = \frac{4\pi R_o^3}{3} \quad \ldots A4.7$$

Equations A4.6, A4.5 and A4.7 =>

$$\rho = \frac{3kc\, R_o^2}{G \times 4\pi R_o^3}$$

$$\rho = \frac{3kc}{G \times 4\pi R_o} \quad \ldots A4.8$$

Equations A4.4 and A4.8 =>

$$M = \int \frac{4\pi r^2\, dr \times 3kc}{4\pi G \times R_o} \quad \ldots A4.9$$

$$M = \int \frac{r^2\, dr \times 3kc}{G \times R_o} \quad \ldots A4.10$$

We know from equation A2.8 that:-

$$a = -kc \quad \ldots A4.11$$

Equations A4.1 and A4.2 and A4.3 and A4.11 =>

$$E = \tfrac{1}{2} M (U^2 - 2kcR_o)\, r/R_o \quad \ldots A4.12$$

Assume $U = c$

Equations A4.10 and A4.12 =>

$$E = \frac{3kc}{2G \times R_o^2} \int r^3 (c^2 - 2kcR_o)\, dr \quad \ldots A4.13$$

$$E = \frac{3kc}{8G \times R_o^2} \left[r^4 (c^2 - 2kcR_o) \right]_0^{R_o}$$

$$E = \frac{3kc \times R_o^2 (c^2 - 2kcR_o)}{8G} \quad \ldots A4.14$$

Equation A4.14 =>

$$E = 0 \text{ when } R_o = 6.59442 \times 10^{25} \text{ m}$$

This coincides exactly with the result from appendix 3 where we calculated the maximum radius of the universe to be the exact radius required for the universe to become a black hole.

CONCLUSION:
This coincides exactly with the result from appendix 3 where we calculated this radius $R_o = 6.59442 \times 10^{25}$ m to be the exact maximum radius of the universe and also the point at which the universe becomes a black hole. This classical calculation is predicting this to be the point at which the total kinetic energy of the universe drops to zero.

Alternative History Of Time

Appendix 5 – Relativistic Kinetic Energy Of Universe

Again, assume the universe mass **M** is expanding. At any point in time the radius of the universe is R_o.

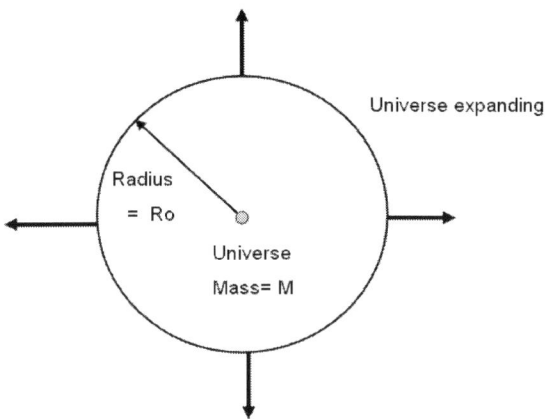

Diagram A5.1 Calculating The Kinetic Energy Of Universe.

For constant acceleration a we know that...

$$V^2 = U^2 + 2aS \qquad \ldots A5.1$$

Where U is the initial velocity and V is the velocity after travelling distance S

If V_o is the velocity of the surface of the universe (at radius R_o) then at any point r ;-

$$V = V_o \times \frac{r}{R_o} \qquad \ldots A5.2$$

Up to this point appendix 5 has been exactly the same as appendix 4 – the calculation of kinetic energy of the universe without taking into account relativity.

However, in the relativistic world we know that the kinetic energy of a moving body is given by;-

$$E = \frac{mc^2}{(1 - v^2/c^2)^{1/2}} - mc^2 \qquad \ldots A5.3$$

The mass of the universe can be calculated as;-

$$M = \int 4\pi r^2 \, dr \times \rho \qquad \ldots A5.4$$

where ρ is assumed to be the constant density of the universe.

Equation 11 =>

$$M = \frac{kcr^2}{G} \qquad \ldots A5.5$$

$$\rho = M/Vol \qquad \ldots A5.6$$

$$Vol = \frac{4}{3}\pi Ro^3 \qquad \ldots A5.7$$

Equations A5.6, A5.5 and A5.7 =>

$$\rho = \frac{3kc\, Ro^2}{G \times 4\pi\, Ro^3}$$

$$\rho = \frac{3kc}{G \times 4\pi\, Ro} \qquad \ldots A5.8$$

where ρ is the *average* density of the universe.

Equations A5.4 and A5.8 =>

$$M = \int \frac{4\pi r^2 \, dr \times 3kc}{4\pi\, G \times Ro} \qquad \ldots A5.9$$

$$M = \int \frac{r^2 \, dr \times 3kc}{G \times Ro} \qquad \ldots A5.10$$

We know from equation A2.8 that;-

$$a = -kc \qquad \ldots A5.11$$

Equations A5.1 and A5.2 and A5.3 and A5.11 =>

$$E = \int \frac{4\pi r^2 \, dr \; \rho\, c^2\, Ro\, c}{(Ro^2 c^2 - Vo^2 r^2)^{1/2}} - mc^2 \qquad \ldots A5.12$$

Using the simple substitution...

$$r = (a/b)^{1/2} \sin \Phi$$

Integrating the right-hand side gives...

$$E = \frac{3kc^4 \times I1}{G} - \frac{kc^3 R_o^2}{G} \qquad \ldots A5.13$$

Where ...

$$I1 = \frac{a}{b^{3/2}} [A - (B \times C)]$$

And ...

$$A = \sin^{-1} \{1 - 2kR_o/c\}^{\frac{1}{2}}$$

$$B = \frac{1}{2} \{1 - 2kR_o/c\}^{\frac{1}{2}}$$

$$C = \{2kR_o/c\}^{\frac{1}{2}}$$

$$a = R_o^2 c^2$$

$$b = c^2 - 2kcR_o$$

This all looks a bit complicated but is quite easily expressed in a simple spreadsheet. There are differences between the calculation of kinetic energy of the expanding universe when relativity is taken into account compared to the case in appendix 4 when it was not taken into account. However as with the case when relativity was not taken into account, the total kinetic energy of the universe drops to zero at the exact radius predicted by the calculation when relativity was not taken into account.

This result is based on the use of an *average* density for the universe which I believe is not entirely valid.

CONCLUSION:
When relativity is taken into account in the expanding universe, the total kinetic energy of the universe drops to zero at exactly the same radius of universe calculated in appendix 4 i.e. **$R_o = 6.59442 \times 10^{25}$ m**

Alternative History Of Time

Appendix 6 – Energy Of Photons Moving To Centre Of Universe.

The acceleration due to gravity decreases linearly inside a uniform gravitational body (for example our universe.) A photon will gain energy as it travels from the outside edges of the universe in towards the centre. Here we calculate just how much energy it gains. The diagram below shows a photon travelling in towards the centre of the universe.

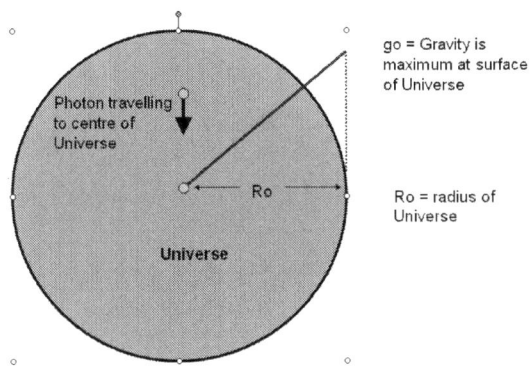

Diagram A6.1. Photons Travelling To The Centre Of The Universe.

How much energy does a photon gain when it moves from the outside edge of a static universe towards the centre?

At the edge of the universe...

$$g_o = \frac{GM}{R_o^2} \quad \ldots A6.1$$

where G is Newton's Gravitational constant and M is the mass of the universe and Ro is the radius of the universe.

The acceleration experienced by the photon decreases linearly. Therefore...

$$a = \frac{r}{R_o} \times \frac{GM}{R_o^2} \quad \ldots A6.2$$

So, we can say ...

$$F = \frac{m \times GM \times r}{R_o^3} \quad \ldots A6.3$$

The force on the photon is:-

$$F = \frac{d(mv)}{dt}$$

$$F = m \times \frac{dv}{dt} + v \times \frac{dm}{dt}$$

But v = c (the speed of light for the photon) so $\frac{dv}{dt} = 0$

Therefore;-

$$F = c \times \frac{dm}{dt} \qquad \ldots A6.4$$

The Energy of the photon is ...

$$E = \int F \times dx \qquad \ldots A6.5$$

Equations A6.4 and A6.5 =>

$$E = \int c \times \frac{dm}{dt} \times dx \qquad \ldots A6.6$$

Equations A6.3 and A6.5 =>

$$E = \int \frac{GMm \, r \, dx}{R_o^3} \qquad \ldots A6.7$$

Equations A6.6 and A6.7 =>

$$\int c \times \frac{dm}{dt} \times dx = \int \frac{GMm \, r \, dx}{R_o^3} \qquad \ldots A6.8$$

Re-arranging equation A6.8 and substituting dx/dt = c we get...

$$\int \frac{c^2}{m} dm = \int \frac{GM \, r}{R_o^3} dx \qquad \ldots A6.9$$

This becomes ...

$$[c^2 \log m]_{m0}^{m1} = \left[\frac{GMr^2}{R_o^3}\right]_{R_o}^{r1}$$

Substituting in limiting values gives...

$$c^2 \log \frac{m_1}{m_o} = \frac{GM}{R_o^3}(R_o^2 - R_1^2) \qquad \text{...A6.10}$$

Therefore, when $R_1 = 0$ i.e. when the photon has reached the centre of the universe...

$$m_1 = m_o \times e \wedge [GM/(R_o \times c^2)] \qquad \text{...A6.11}$$

$$E = mc^2 = hf = hc/\lambda$$

So ...

$$\lambda = \frac{h}{mc} \qquad \text{...A6.12}$$

Equations A6.11 and A6.12 =>

$$\lambda_1 = \lambda_o / e \wedge [GM/(R_o \times c^2)] \qquad \text{...A6.13}$$

When the universe is at its maximum expansion, we showed in appendix 3 that it just becomes a black hole.

For an object just that just becomes a black hole, we know from equation A3.7 that...

$$[GM/(R_o \times c^2)] = 0.5$$

Therefore, equation A6.13=>

$$\lambda_1 = \lambda_o / e^{0.5}$$

or
$$f_1 = f_o \times 1.6487 \qquad \text{...A6.14}$$

Where f_1 is the frequency of light at the centre of the universe at maximum extent and f_0 is the frequency on the edge of the universe.

Alternative History Of Time

Appendix 7 – Variable Density Of The Universe

The density of the universe ρ_r at distance r from the place where the 'Big Bang' occurred can be calculated.

We know from the alternative interpretation of Hubble's law described earlier, that since the 'Big Bang', that there has been a constant deceleration of ;-

$$a = -kc \quad \ldots A7.1$$

where k is Hubble's constant and c is the speed of light.

But we know from Newton's law of Gravitation that…

$$a = -GM/r^2 \quad \ldots A7.2$$

where G is Newton's Gravitational constant and M is the mass of the universe with radius r.

Combining A7.1 and A7.2 we can say…

$$M = kc \times r^2/G \quad \ldots A7.3$$

If the density of any portion of the universe is inversely proportional to the distance from the centre we can say;-

$$\rho_r = B/r \quad \ldots A7.4$$

where B is a constant. *We know this to be the case from the following consideration…*

The mass of the universe can be calculated from the density;-

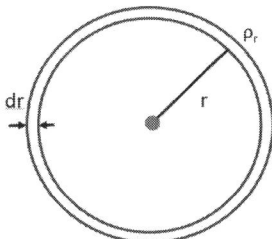

Diagram A7.1 The Density of the universe at distance r from the centre.

$$M = \int \rho_r \times 4\pi r^2 \, dr \quad \ldots A7.5$$

Here we can see ρ_r has to be inversely proportional to radius r otherwise the integration shown above wouldn't produce a relationship for the mass of the universe which is directly proportional to the radius squared as shown in equation A7.3!

A7.4 & A7.5 =>

$$M = \int B/r \times 4\pi r^2 \, dr \quad \ldots A7.6$$

Therefore...

$$M = 2\pi B r^2 \quad \ldots A7.7$$

Comparing this result to equation A7.3 we can say;-

$$B = kc/2\pi G \quad \ldots A7.8$$

Therefore, we can conclude that the density of the universe ρ_r at distance r from the centre is given by the expression;-

$$\rho_r = kc/(2\pi G r) \quad \ldots A7.9$$

where k is Hubble's constant, c is the speed of light and G is Newton's Gravitational constant.

Note: ρ_r is the density at the very distance r, *not the average density of the universe of radius r.*

Equation A7.9 can be written as;

$$\rho_r = 1.6252/r \quad \ldots A7.10$$

Thus, we can see that the density of the growing universe would have reduced quite rapidly after the initial 'Big Bang'. At a distance of just one metre from the very centre of the universe, the density of that region of space would only be about 1.6252 kg/m^3 which is less than the density of some gases we encounter here on Earth at standard temperature and pressure (e.g. carbon dioxide 1.977 kg/m^3).

We can use this result to calculate the average distance between stars for our region of space at a distance of 24 million light years from the site of our 'Big Bang' event. Assuming the sun to have an average star mass, this calculated average distance between stars comes out to be about 30 light years (somewhat larger than the 5 light year distance to our nearest star neighbour, although this is not necessarily significant as we can expect large regional variations away from the average as stars do tend to cluster.)

At the maximum radius (6.97 billion light years) which the universe will eventually reach when it stops expanding ($r_{max} = c/2k$ where k is Hubble's constant) the calculated average distance between stars comes out to be about 196 light years again using equation A7.10 to calculate the average density at a distance r from the centre of the universe and assuming the average star mass is equivalent to the mass of the sun.

Average Density of Universe

The overall average density of the universe ρ_{av} taking into account the entire mass of the universe and its entire volume is given by;-

$$\rho_{av} = M/\text{Volume} \qquad \ldots A7.11$$

Equations A7.3 & A7.11 =>

$$\rho_{av} = 3kc/4\pi Gr \qquad \ldots A7.12$$

Equations A7.9 & A7.12 =>

$$\rho_{av}/\rho_r = 3/2 \qquad \ldots A7.13$$

Therefore, throughout its lifetime, the average density of the universe is 1.5 times larger than the minimum density seen at the outer edges.

The velocity V at which the outer edge of the universe of radius r moves outwards is given by;-

$$V^2 = U^2 - 2ar \qquad \ldots A7.14$$

where $U = c$ and $a = kc$ where k is Hubble's constant.

The maximum radius is reached when $V = 0$

Therefore;-

$$r_{max} = c/2k \qquad \ldots A7.15$$

Equations A7.9 and A7.15 =>

$$\rho_{min} = k^2/\pi G \qquad \ldots A7.16$$

This equates to 2.46×10^{-26} kg/m^3 equivalent to about one atom for every two cubic metres or about half the density of the Inter Galactic Medium (IGM) observed in our region of space.

Alternative History Of Time

Appendix 8 – Mass Extinctions

It might be the case that some of the various mass extinctions that we are aware of, were caused by "aftershocks" from the original 'Big Bang'. This is not known for certain, but in this section, I theorise about the possible timing of any such events.

Whenever there is an incident like a large volcanic eruption or a major earthquake or any similar massive impact event, it is quite often the case that further disruptions occur sometime after the primary event has completed. Similar types of disruptions might have occurred after the initial 'Big Bang'. Here I calculate the possible timings of any such aftershocks that must have occurred, if they were indeed to be held responsible for any subsequent mass extinctions.

Let us assume that the 'Big Bang' released a massive amount of electromagnetic radiation which travels at the speed of light 'c' and bounced off the boundary of the spherical universe.

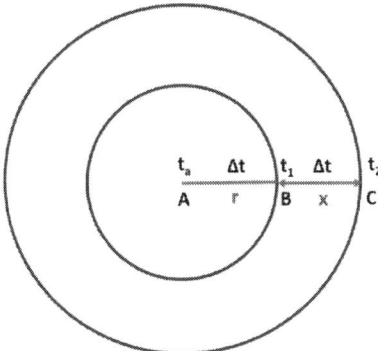

Diagram A8.1 – Radiation emitted from the centre of the universe.

Assume the radiation leaves the centre of the universe from point A at time t_a It travels at speed 'c' for time Δt and bounces off the boundary of the universe at point B. At this time the universe has expanded to radius 'r' in time t_1

So, we can say...

$$r = \Delta t/c \qquad \text{... A8.1}$$

and

$$t_1 = t_a + \Delta t \qquad \text{... A8.2}$$

The radiation then travels back from point B back to the centre of the universe at point B and takes time Δt to travel back from point B to point A.

Alternative History Of Time

During this time Δt, the radius of the universe increases by distance x and therefore reaches radius r + x by the time the universe has reached an age of t_2

So, we can say;-

$$t_2 = t_1 + \Delta t \qquad \text{... A8.3}$$

As already discussed, there is a strong argument to say that the Milky Way galaxy is located very close to point 'A' at the centre of the universe. So, the time t_2 is given by the age of the universe when extinctions caused by this radiation occurred.

It seems quite possible that four of the last five mass extinctions which occurred on Earth could have resulted from this type of radiation. It might be the case that this has occurred indirectly – for example, large meteorites being generated by significant disturbance to stars including our own Sun.

The table below shows when the last five major extinctions have occurred so we can calculate the timings of the aftershocks that would have occurred, if indeed these extinctions resulted from echoes of the 'Big Bang' as postulated here.

Of these mass extinctions, it seems likely that the Triassic-Jurassic Extinction which occurred 210 million years ago does not fit the sequence of extinctions caused by 'Big Bang' echoes as it occurred too soon after the Permian-Triassic Extinction. It's assumed the others extinctions listed here resulted from echoes of the 'Big Bang'.

Ref	Extinction Event	Million years ago
1	Ordovician-silurian Extinction	440
2	Devonian Extinction	365
3	Permian-triassic Extinction	250
4	Triassic-jurassic Extinction	210
5	Cretaceous-tertiary Extinction	65

Table A8.1 Major Extinctions

We can use these results to calculate the values of t_a the times when the 'Big Bang' radiation must have passed though the centre of the universe.

We know from the alternative interpretation of Hubble's law that the universe is undergoing constant deceleration a = -kc

Since

$$s = ut + 1/2\, at^2 \qquad \text{... A8.4}$$

$$r = ct_1 - 0.5kct_1^2 \qquad \text{... A8.5}$$

Using A8.1 =>

$$r = c \times \Delta t \qquad \ldots A8.6$$

Putting this into A8.5 =>

$$c \times \Delta t = ct_1 - 0.5kct_1^2 \qquad \ldots A8.7$$

but A8.3 =>

$$\Delta t = t_2 - t_1 \qquad \ldots A8.8$$

Therefore A8.7 and A8.8 =>

$$c(t_2 - t_1) = ct_1 - 0.5kct_1^2 \qquad \ldots A8.9$$

Solving for t_1 yields...

$$t_1 = 2/k[1 +/- \{1 - kt_2/2\}^{0.5}] \qquad \ldots A8.10$$

So, we can use this formula and the values of t_2 when the extinctions occurred to calculate values for t_1

Then equation A8.8 can be used to calculate the value of Δt and equation A8.2 can then be used to calculate the equivalent values for t_a. The results are shown below;-

Extinctions	Years Ago	t_2 Age of Universe at Extinction/yrs	t_2 Age of Universe at Extinction/secs	t_1/secs	$\Delta t = r/c$	$t_a = t_1 - \Delta t$
10	4.40E+08	1.3849E+10	4.3675E+17	2.5546E+17	1.8129E+17	7.4167E+16
11	3.65E+08	1.3924E+10	4.3911E+17	2.5712E+17	1.8199E+17	7.5139E+16
12	2.50E+08	1.4039E+10	4.4274E+17	2.5969E+17	1.8305E+17	7.6647E+16
13	6.50E+07	1.4224E+10	4.4857E+17	2.6384E+17	1.8473E+17	7.9118E+16

Table A8.2 Calculated emissions from centre of universe t_a

The current age of the universe used in this calculation was 14.289 billion years which is 3.7% higher than the normally accepted value. Also, the occurrence of the most recent extinction caused by 'Big Bang' echoes was found by trial and error to be number 13. The aim was to get the resulting graph of t_a v N to pass through zero. The resulting graph is shown below.

This graph misses the zero point by 643,500 seconds which might sound like a lot, but this equates to about 7.4 days which is fairly close to zero when compared to a 14-billion-year lifetime of the universe.

Alternative History Of Time

The graph shows the relationship between t_a the time at which the echo of the 'Big Bang' passed through the centre of the universe and the order of each resulting mass extinction on Earth.

We see;-

$$t_a = 7.115 \times 10^{13} N^3 - 2.080 \times 10^{15} N^2 + 2.110 \times 10^{16} N - 6.435 \times 10^5 \quad \ldots A8.11$$

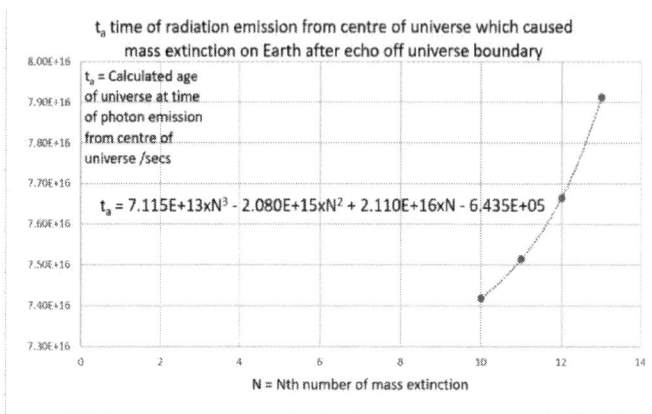

Graph A8.1 – 'Big Bang' echoes through centre of universe t_a

The formula for t_a shown on the graph can be used to calculate timings t_2 for all extinctions...

Since;
$$t_1 = t_a + \Delta t \quad \ldots A8.12$$

Then
$$\Delta t = t_1 - t_a \quad \ldots A8.13$$

And
$$r = c \times \Delta t \quad \ldots A8.14$$

Since
$$r = ct_1 - 1/2 kct_1^2 \quad \ldots A8.15$$

Combining A8.13, A8.14 and A8.15 we can say...

$$c(t_1 - t_a) = ct_1 - 1/2 kct_1^2 \quad \ldots A8.16$$

Therefore
$$ct_a = 1/2 kct_1^2 \quad \ldots A8.17$$

So ...

$$t_1 = (2t_a/k)^{0.5} \qquad \ldots A8.18$$

Having calculated t_1 from t_a it's easy to calculate t_2

Equation A8.10 can be re-arranged to give;

$$t_2 = 2/k[1 - \{ kt_1/2 - 1\}^{0.5}] \qquad \ldots A8.19$$

The results of these calculations are shown in the table below;

N	N^2	N^3	t_a calculated back/secs	$t_1 = (2t_a/k)^{0.5}$	$r = ct_1 - 0.5kct_1^2$	$\Delta t = r/c$	$t_2 = t_1 + \Delta t$	t_2/yrs	(Current age universe - t_2)/million yrs	t_2/current age of universe	Actual Extinctions/ millions years ago
1	1	1	1.91E+16	1.296E+17	3.313E+25	1.105E+17	2.401E+17	7.614E+09	6,675	53.287%	
2	4	8	3.44E+16	1.741E+17	4.187E+25	1.397E+17	3.138E+17	9.949E+09	4,340	69.627%	
3	9	27	4.65E+16	2.023E+17	4.670E+25	1.558E+17	3.580E+17	1.135E+10	2,935	79.457%	
4	16	64	5.57E+16	2.213E+17	4.966E+25	1.657E+17	3.870E+17	1.227E+10	2,018	85.877%	
5	25	125	6.24E+16	2.343E+17	5.154E+25	1.719E+17	4.062E+17	1.288E+10	1,408	90.146%	
6	36	216	6.71E+16	2.430E+17	5.273E+25	1.759E+17	4.188E+17	1.328E+10	1,008	92.945%	
7	49	343	7.02E+16	2.485E+17	5.346E+25	1.783E+17	4.268E+17	1.353E+10	755	94.718%	
8	64	512	7.21E+16	2.519E+17	5.390E+25	1.798E+17	4.317E+17	1.369E+10	601	95.793%	
9	81	729	7.33E+16	2.539E+17	5.416E+25	1.807E+17	4.346E+17	1.378E+10	508	96.442%	
10	100	1000	7.42E+16	2.554E+17	5.435E+25	1.813E+17	4.367E+17	1.385E+10	441	96.912%	440
11	121	1331	7.51E+16	2.571E+17	5.455E+25	1.820E+17	4.391E+17	1.392E+10	366	97.436%	365
12	144	1728	7.66E+16	2.597E+17	5.487E+25	1.830E+17	4.427E+17	1.404E+10	251	98.240%	250
13	169	2197	7.91E+16	2.638E+17	5.538E+25	1.847E+17	4.485E+17	1.422E+10	67	99.534%	65
14	196	2744	8.30E+16	2.702E+17	5.612E+25	1.872E+17	4.574E+17	1.450E+10	-214	101.500%	
15	225	3375	8.86E+16	2.793E+17	5.715E+25	1.906E+17	4.699E+17	1.490E+10	-611	104.275%	

Table A8.3 Calculated times t_2 of extinctions caused by 'Big Bang' echoes

It can be seen that the predictions for the 10th, 11th and 12th extinctions agree with actual events to within 1 million years and the calculation for the 13th event is accurate to within 2 million years. It would be interesting to establish how well all the calculated past extinctions listed here match with archaeological evidence.

The next future such event is predicted to occur in 214 million years from now or thereabouts, depending on the rate at which our Milky Way galaxy is moving further away from the centre of the universe.

Although the next mass extinction is a long way into the future, this would of course not be the case for any galaxy located 214 million light years away from the centre of the universe. They would *currently* be suffering a mass extinction event.

These estimates are dependent on the many assumptions made in this calculation. In particular, the actual geometry of the universe might not be a true sphere and this might therefore impact the timings of the reflections of 'Big Bang' energy.

Alternative History Of Time

Appendix 9 – Average Temperature Of The Universe

In appendix 7 the density of the universe ρ_r at distance r from the centre was calculated. From equation A7.9 it was deduced;-

$$\rho_r = kc/(2\pi Gr) \qquad \ldots A9.1$$

This can be used to calculated the kinetic energy of the universe.

Assume the maximum radius of the universe at any time is R_o and the velocity of the boundary of the universe at distance R_o from the centre is V_o

We can say the velocity at any distance r is given by;

$$V_r = V_o \times r/R_o \qquad \ldots A9.2$$

Kinetic Energy is calculated from;-

$$KE = \tfrac{1}{2} mV^2 \qquad \ldots A9.3$$

Therefore, we can say;-

$$KE = \int 0.5 \times \rho_r \times 4\pi r^2 \, dr \times (V_o \times r/R_o)^2 \qquad \ldots A9.4$$

Substituting for ρ_r from equation 9.1 and integrating gives;-

$$KE = [k_h c \times V_o^2 \times R_o^2 /4G] \qquad \ldots A9.5$$

The mass of the entire universe M was previously derived in equation A7.3 =>

$$M = k_h c \times R_o^2/G \qquad \ldots A9.6$$

Therefore, the *average* Kinetic Energy per unit mass is given by;-

$$KE/M = 0.25 \times V_o^2 \quad \ldots A9.7$$

We can use KE/M = SHC x T to estimate the average temperature T of the universe where SHC is assumed to be the Specific Heat Capacity of hydrogen 14,300 J/kg.K I've chosen the value for hydrogen as this is thought to be the most common element in the universe. The Specific Heat Capacity of hydrogen varies significantly at different temperatures and pressures, but by using this value for estimation purposes, we find that the *average* temperature of the universe will not fall to a level of about 300 Kelvin (~27°C thought to be a reasonable temperature to support life) until the universe is just 267,000 years from reaching its maximum radius. This will happen when the radius of the universe is just 2.55 light years from its maximum size. It is thought that at the present time the universe is about 150 million years from reaching its maximum radius and that the average temperature across the universe at the present time is still about 186 million degrees Celsius.

Thus, it's believed that, only as the expansion of the universe comes to an end, will the temperature of the *majority* of the universe become conducive to support life.

This again illustrates how fortunate we appear to be to live in a quiet neighbourhood located in space where the temperature is so modest.

Appendix 10 – Estimating The Initial Mass of the Universe

In appendix 9 the average Kinetic Energy per mass of the universe was calculated.

From equation A9.10 it was deduced;-

$$KE/M = \mathbf{0.25} \times V_o^2 \qquad \ldots A10.1$$

where V_o is the velocity of the outer edge of the universe.

But we can express the ratio of KE/M as;

$$KE/M = SHC \times \Delta T \qquad \ldots A10.2$$

where SHC is the Specific Heat Capacity of the material in the universe and ΔT is the temperature of that material.

We can express energy in different ways.

$$E = 3/2 \times K_B T \qquad \ldots A10.3$$

where K_B is Boltzman's constant.

We can also say;-

$$E = hf \qquad \ldots A10.4$$

where h is Planck's constant and f is the frequency of light that would consist of photons of energy E.

Combining equations A10.3 and A10.4 we can say;-

$$T = 2hf/(3 \times K_B) \qquad \ldots A10.5$$

Putting this expression for T in equation A10.2 gives;-

$$KE/M = SHC \times 2hf/(3 \times K_B) \qquad \ldots A10.6$$

A10.1 and A10.6 =>

$$V_o^2/4 = SHC \times 2hf/(3 \times K_B) \qquad \ldots A10.7$$

But we know;-

$$V_o^2 = C^2 - 2ar \qquad \ldots A10.8$$

Alternative History Of Time

where $a = K_H C$ and r is the radius of the universe.

A10.7 and A10.8 =>

$$C^2 - 2 K_H C \times r = SHC \times 8hf/(3 \times K_B) \qquad \text{...A10.9}$$

Therefore...

$$f = 3K_B \times C^2/(8h \times SHC) - 3K_H \times C \times r \times K_B/(4h \times SHC) \qquad \text{...A10.10}$$

When $r = 0$ then;-

$$f = 3K_B \times C^2/(8h \times SHC) \qquad \text{...A10.11}$$

Since the energy equivalent of a mass is;-

$$E = mc^2 \qquad \text{...A10.12}$$

and 10.4 => $E = hf$ therefore the mass of a particle can be expressed in terms of the frequency f;-

$$m = hf/c^2 \qquad \text{...A10.13}$$

Substituting from equation A10.11 for f when the radius of the universe is $r = 0$ then;-

$$M_o = 3K_B/(8 \times SHC) \qquad \text{...A10.14}$$

where M_o is the initial mass of the universe.

If we assume the Specific Heat Capacity to be roughly equivalent to hydrogen then SHC = 14,304 J/Kg K

It's difficult to know what the initial Specific Heat Capacity of the universe would be. As far as we're aware, hydrogen appears to be the most common element in the universe, but the value of the Specific Heat Capacity of hydrogen does vary significantly depending on the conditions such as temperature and pressure.

Using this value for SHC in equation 10.14 yields an estimate for the initial mass of the universe =>

$$M_o = 3.62 \times 10^{-28} \text{ kg}$$

This is equivalent to the mass of 397 electrons or $0.216 \times M_p$ where M_p is the mass of a proton.

There is no known fundamental particle with this exact mass. The muon (mu lepton) has a mass of 206.85 electron masses.

If a mass of this size were converted to energy, it might create a photon of frequency f given by equation A10.13 =>

$$f = mc^2/h \qquad \ldots A10.15$$

And a wavelength of ...

$$\lambda = h/mc \qquad \ldots A10.16$$

These equations yield the values;-

$$f = 4.91 \times 10^{22} \text{ Hz} \qquad \ldots A10.17$$

$$\lambda = 6.11 \times 10^{-15} \text{ metres} \qquad \ldots A10.18$$

If we use equation A9.6 to calculate the radius of the universe R when it has the mass of 397 electrons we find;-

$$R = 5.95 \times 10^{-15} \text{ metres} \qquad \ldots A10.19$$

The difference between the calculated wavelength and this radius of the 397 electron universe is less than 3%

It seems possible that a shrinking universe would explode in another 'Big Bang' once its size had decreased to such a small size.

What could make a tiny universe so unstable?

The reason a small universe might become unstable remains a mystery. It might be explained by the Uncertainty principle.

We know that the product of the uncertainty in energy ΔE and the uncertainty in time Δt must always be greater than $h/4\pi$ where h is Planck's constant.

$$\Delta E \times \Delta t > h/4\pi \qquad \ldots A10.20$$

where $\quad \Delta E = mc^2 \qquad \ldots A10.21$

and $\quad \Delta t = r/c \qquad \ldots A10.22$

Therefore, if the product m x c x r of the universe reached the value $h/4\pi$ it would become inherently unstable. We can calculate the radius of the universe when this might occur.

Alternative History Of Time

From equations A10.21 and A10.22 we can say the critical radius is reached when...

$$m \times c \times r = h/4\pi \qquad \text{...A10.23}$$

Since equation A9.6 gives the mass of the universe in terms of its radius...

$$M = k_h c \times r^2/G \qquad \text{...A10.24}$$

We can combine A10.23 and A10.24 to give the critical radius...

$$r = (hG/4\pi K_h c^2)^{1/3} \qquad \text{...A10.25}$$

The resulting radius $r = 2.5825 \times 10^{-15}$ metres giving an equivalent mass of 74.8 electrons so somewhat below the 397-electron limit calculated above. It might be the case that the tiny superheated universe would reach some other boundary condition before the limit of the Uncertainty principle could be reached.

Appendix 11 – Observing An Expanding Universe

For reasons already explained, I think we are extremely fortunate to be located very near to the relatively calm centre of the universe, where the ambient temperature has reduced sufficiently to allow life to evolve. It's perhaps an interesting thought to wonder what the boundary of the expanding universe might look like and to wonder if we'd ever be able to observe it and find out what might lie beyond it.

Unfortunately, it will never be possible to view the true current state of the boundary of our *expanding* universe for much the same reason that was explained earlier in relation to my alternative interpretation of Hubble's law. Given the finite velocity of light, we will only ever be able to observe the boundary of the universe as it was at some time in the distant past. The extent of this limitation is explained here.

As stated earlier in this text, I've assumed that the initial expansion of the universe began at the point of the 'Big Bang' and at that time the boundary of the universe travelled outwards at a speed very close to the maximum permissible speed in our universe, which is c the speed of light.

However, according to my alternative interpretation of Hubble's law, immediately after the 'Big Bang', the boundary of the universe experienced constant deceleration of $a = -k_H c$ (where k_H is Hubble's constant and c is the speed of light.)

Suppose we observe the maximum radius of the universe r_1 as it was at some point in the past when the universe had reached the age of t_1 and the speed of the boundary of the universe had slowed to V

Assuming we are located at some point A at the centre of the universe, at the point where the 'Big Bang' initially occurred, the light from the boundary at point B would take some time to reach us. We might receive this light when the age of the universe is t_2

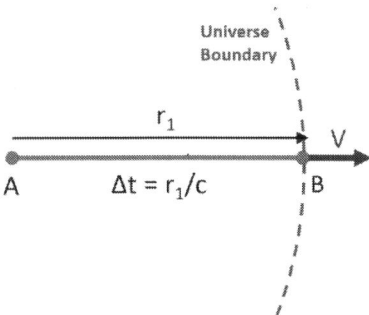

Diagram A11.1 – Observing An Expanding Universe

The light from the boundary would take time r_1/c to reach us.

Therefore, we could say;-

$$t_2 = t_1 + r_1/c \qquad \text{... A11.1}$$

Alternative History Of Time

Assuming the initial speed of expansion of the universe occurred at the speed of light c then it would only have been possible (if such a thing could ever have been possible) for any theoretical observer standing at the centre of the universe to observe the boundary of the universe as it was at some point very close to half the age of the universe, given that light would travel from the boundary of the universe back to the observer at much the same speed the boundary of the universe would be moving away.

But as time passed and the universe aged, the boundary of the universe would have experienced constant Hubble deceleration and the expansion of the universe would therefore have slowed down.

Assuming the expansion of the universe is still ongoing (and calculations suggest that this is probably the case) then it would be theoretically possible for an observer standing at the centre of the universe to see beyond the radius the universe had reached at half the age it had reached.

Using the well-known formula for calculating distance 's' in the case of constant acceleration 'a'

$$s = ut + 1/2 \, at^2 \qquad \ldots \text{A11.2}$$

We can say;-

$$r = ct_1 - 1/2 \, kct_1^2 \qquad \ldots \text{A11.3}$$

where k is Hubble's constant.

Putting this into equation A11.1 gives;-

$$1/2 \, kt_1^2 - 2t_1 + t_2 = 0 \qquad \ldots \text{A11.4}$$

Solving this quadratic equation for t_1 gives...

$$t_1 = \{2 +/- (4 - 2kt_2)^{0.5}\}/k \qquad \ldots \text{A11.5}$$

We can calculate various ages t_1 for the age of the boundary of the universe that theoretically we might be able to observe as the age of the universe t_2 increases.

If we plot these results, we get the graph shown below;-

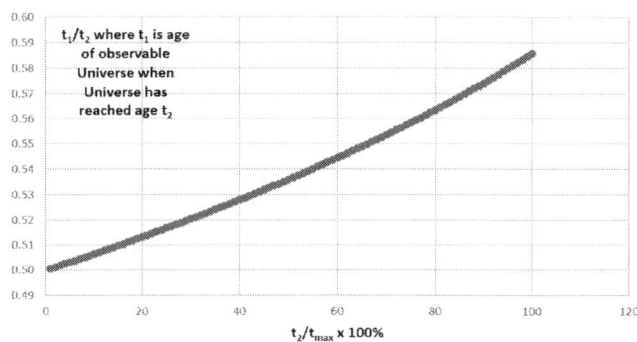

Graph A11.1 – The Observable universe increases as the expansion slows

Since with constant acceleration, the velocity of the boundary of the universe V is given by;

$$V = U + at$$

where U is the initial speed of the boundary (U = c)

We can say when V = 0 then the radius of the universe would be a maximum.

Therefore, the value of t_2 would be...

$$t_2 = 1/k \qquad \text{... A11.6}$$

So, when the universe has reached its maximum size and stopped expanding then;

$$t_1 = [2 +/- 2^{0.5}]/k \qquad \text{... A11.7}$$

$$t_1 = 0.586 \times t_2 \qquad \text{... A11.8}$$

It's generally thought that the age of the universe at the present time is $t_2 = 13.8$ billion years

i.e.
$$t_2 = 13,800,000,000 \text{ years}$$

If we increase this age by just **one year** then the value of t_2 increases to;-

$$t_2 = 13,800,000,001 \text{ years}$$

Alternative History Of Time

With this increase of just one year in the current age of the universe t_2, the observable age t_1 increases by 256.7 days. So, every year that passes by, we get to see that little bit more of the universe.

t_2/t_{max}	t_2/secs	t_1/secs	t_1/t_2	t_2/years	$r_1 = ct_1 - 0.5kct_1^2$	r_1/r_{max}
96.00	4.223E+17	2.454E+17	0.581	1.339E+10	5.305E+25	0.8044
97.00	4.267E+17	2.484E+17	0.582	1.353E+10	5.345E+25	0.8105
98.92	4.352E+17	2.544E+17	0.584	1.380E+10	5.421E+25	0.8221
99.00	4.355E+17	2.546E+17	0.585	1.381E+10	5.424E+25	0.8225
100.00	4.399E+17	2.577E+17	0.586	1.395E+10	5.463E+25	0.8284

Table A11.1 – Showing the viewable radius of the universe

The table above shows the theoretical viewable radius of the universe. At the present time we should theoretically be able to see 82.21% of the maximum radius the universe will eventually reach but eventually, at the exact point the universe stops expanding, anybody still living on Earth in about 150 million years in the future ought to be able to receive light to observe 82.84% of the maximum radius of the universe.

*** THE END ***

P.J. Naughton

Alternative History Of Time

P.J. Naughton

Printed in Dunstable, United Kingdom